Internet of Things (IoT) Its Applications

Dr. V K Sachan
B.Tech (Hons.), M.Tech (Hons.), Ph.D.
Professor

Copyright @ 2020, Smt. Jay Devi Sachan Memorial Publication House

All rights reserved. No part of this publication may be reproduced or copied as any material without the written permission of the authors. Photocopying in it or storing it in any medium as graphics, electronics, mechanical means and for the use of something is not transitory or accidental. Any violation of this will be subject to legal action and prosecution without notice.

First Edition: 2020

ISBN: 9798656012751

Published by

Smt. Jay Devi Sachan Memorial Publication House

India
Email ID: drvksachan8@gmail.com

Preface

IoT is the biggest opportunity ever for our industry. With capabilities much greater than today's networks, opportunities beyond our imagination will appear. With IoT, we will be able to digitalize industries and realize the full potential of a networked society. The Internet of Things (IoT) is a system of interrelated computing devices, mechanical and digital machines, objects, animals or people that are provided with unique identifiers and the ability to transfer data over a network without requiring human-to-human or human-to-computer interaction. The book also looks at all the sub-systems of the Internet of Things, focusing on both the practical and theoretical issues. This text book "Internet of Things (IoT) & Its Applications" is organized into Ten Chapters.

Chapter-1: Introduction to IoT for Beginners
Chapter-2: Building Blocks of Internet of Things (IoT) and Their Characteristics
Chapter-3: Domain specific IoT and Their Real-world Applications
Chapter-4: Sensor and Actuator
Chapter -5: Generic Design Methodology and an IoT System Management
Chapter-6: Multiple Protocols in IoT Domain
Chapter-7: Common Security Measures used for Designing an IoT Applications
Chapter-8: Python Logical Design of an IoT system
Chapter -9: Laboratory Companion for Designing IoT Applications
Chapter-10: Challenges and Future Scope of IoT

Salient Features
- Comprehensive Coverage of Basics of IoT, Domain Specific IoT and their real world Applications, Sensor and Actuators, and Generic Design Methodology and IoT System Management.
- New elements in book include Multiple Protocols in IoT Domain, Python Logical Design of an IoT system, Laboratory Companion for Designing IoT Applications and Common Security Measures used for Designing an IoT Applications.
- Clear perception of the various project with Python Logical Design of an IoT system and illustrative diagrams.
- Simple Language, easy- to- understand manner.

Our sincere thanks are due to all Scientists, Engineers, Authors and Publishers, whose works and text have been the source of enlightenment, inspiration and guidance to us in presenting this small book. I will appreciate any suggestions from students and faculty members alike so that we can strive to make the text book more useful in the edition to come.

Dr. V. K. Sachan

Index

Chapter-1: Introduction to IoT for Beginners ... 9
1.1 Introduction
1.2 Internet of Things (IoT)
1.2.1 Why IoT Matters
1.2.2 Examples of IOT
1.2.3 Advantages of IOT
1.2.4 Disadvantages
1.3 Integration of Mobile Apps with IOT
1.3.1 IPv6
1.3.2 Applications
1.4 Working of IoT
1.4.1 Challenges with the Internet of Things

Chapter-2: Building Blocks of Internet of Things (IoT) and Their Characteristics ... 18
2.1 Trends and Characteristics
2.1.1 Architecture
2.1.2 Network Architecture
2.1.3 Application Layer
2.2 IoT Architecture
2.2.1 IoT Architecture Technology
2.2.2 Working of IoT
2.3 Functions of 4 Stage Architecture of IoT
2.4.3 Hardwares used in IOT
2.5 Design Steps involved in a Basic IoT System

Chapter-3: Domain specific IoT and Their Real-world Applications ... 33
3.1   Introduction
3.1.1 Connect with Things
3.1.2 Search for Things
3.1.3 Manage Things
3.1.4 IoT stands Today
3.2   Applications of IoT
3.2.1 Smart Homes
3.2.2 Home Automation and Smart Cities

3.2.3 Environment Monitoring
3.2.4 Medical and Healthcare
3.2.5 Energy Management and Smart Grid
3.2.6 Future of Banking
3.2.7 Retail
3.2.8 Transportation
3.2.9 Media and Advertising
3.2.10 Manufacturing, Oil, Gas and Mining
3.2.11 Construction
3.2.12 Agriculture
3.2.13 Fitness
3.2.14 Security
3.2.15 Wearable Technology
3.2.16 Smart Cars
3.2.17 Commercial Application
3.2.18 Consumer Application
3.2.19 Infrastructure Application
3.2.20 Industrial Application

Chapter-4: Sensor and Actuator ... 42

4.1 Introduction
4.1.1 Characteristics of Good Sensor
4.2 Input and Output System using Sound Transducers
4.2.1 Input Devices or Sensors
4.2.2 Output Devices or Actuators
4.2.3 Types of Sensors
4.2.4 Analogue and Digital Sensors
4.2.5 Signal Conditioning of Sensors
4.3 Capacitive Displacement Sensor
4.3.1 Basic Capacitive Theory
4.3.2 Advantages and Disadvantages of Capacitive Transducer
4.3.3 Applications of Capacitive Transducer
4.4 Piezoelectric Sensor
4.4.1 Piezoelectric Effect
4.4.2 Applications of Piezoelectric Sensor
4.4.3 Principle of Operation
4.4.4 Electrical Properties
4.4.5 Sensor Design
4.4.6 Materials used for the Piezoelectric Transducers

4.4.7 Advantages of Piezoelectric Transducers
4.4.8 Limitations of Piezoelectric Transducers
4.4.9 Applications of the Piezoelectric Transducers
4.5 Hall Effect Sensor
4.5.1 Hall Probe
4.5.2 Hall Effect Sensor Principals
4.5.3 Hall Effect Magnetic Sensor
4.5.4 Materials for Hall Effect Sensors
4.5.5 Signal Processing and Interface
4.5.6 Advantages
4.5.7 Disadvantages
4.5.8 Applications of Hall Effect Sensor
4.6 Photoelectric Sensor
4.6.1 Types of Photoelectric Sensor
4.6.2 Sensing Modes
4.6.3 Photoelectric Effect
4.6.4 Emission Mechanism
4.6.5 Experimental Observations of Photoelectric Emission
4.7 Light Sensors
4.7.1 The Photoconductive Cell
4.7.2 Photo Junction Devices
4.7.3 The Phototransistor
4.7.4 Photovoltaic Cells
4.8 Temperature Sensors
4.8.1 The Thermostat
4.8.2 The Thermistor
4.8.3 Resistive Temperature Detectors (RTD)
4.8.4 The Thermocouple
4.9 Position Sensors
4.9.1 The Potentiometer
4.9.2 Inductive Position Sensors
4.9.2.1 Linear Variable Differential Transformer
4.9.2.3 Rotary Encoders
4.10 Force Sensors
4.11 Motion Sensors
4.12 Fluid Sensors
4.13 Environmental Sensors

Chapter -5: Generic Design Methodology and an IoT System Management ... 101
5.1 Technical Standards of IoT
5.2 Generic Design Methodology
5.4 IoT Project
5.3 IoT System Management
5.4.1 The challenges of Developing an IoT Project
5.4.2 Design Steps for Developing an IoT Project

Chapter-6: Multiple Protocols in IoT Domain ... 107
6.1 Architecture of IoT
6.2 IoT Architecture
6.2.1 The Fundamental Layers of IoT Architecture
6.2.2 Key Components of IoT
6.3 Basics of IoT Architecture
6.3.1 IoT Architecture Layers
6.4 **Common Issues and Threats of** IoT Architecture Layers
6.4.1 Application Layer
6.4.2 Data Processing Layer
6.4.3 Network Layer
6.4.4 Perception layer/Sensor layer

Chapter-7: Common Security Measures used for Designing an IoT Applications ... 117
7.1 Introduction
7.2 IoT Matters
7.2.1 Collecting and Sending Information
7.2.2 Receiving and Acting on Information
7.3.3 Doing Both
7.3 Common IoT Security Measures
7.3.1 IoT Examples
7.3.2 Technology used in IoT
7.3.3 Testing IoT
7.4 IoT Testing Challenges
7.5 IoT Product Developments
7.6 Steps for IoT Product Developments

Chapter-8: Python Logical Design of an IoT system ... 125

8.1 Python in Internet of Things (IoT)
8.2 Python
8.2.1 Python in IoT Development
8.3 Advantages of Python for IoT
8.3.1 Raspberry Pi for Python in IoT
8.4 Project Based on Measurement of temperature and humidity Using IoT
8.4.1 Requirements of Things
8.4.2 Example of Python Code
8.5 Project Based on ON/OFF LED or Interfacing Temperature Sensor or Speed Control of Motor Using IoT
8.6 An IoT project to read a temperature sensor and send data to the cloud

Chapter -9: Laboratory Companion for Designing IoT Applications ... 131

9.1 Internet of Things (IoT)
9.2 Advanced IoT
9.2.1 Internet of Things Learning Outcomes
9.3 Arduino
9.3.1 Important Features of Arduino Board
9.4 Raspberry Pi based IoT Projects
9.4.1 Raspberry Pi IoT based Smart Energy Monitor
9.4.2 IoT based Smart Parking System
9.4.3 IoT based Smart Irrigation System

Chapter-10: Challenges and Future Scope of IoT ... 157

10.1 Challenges in IoT
10.1.1 Major Challenges in IoT
10.2 Future Scope of IoT
10.2.1 Addiction to Tech Connections
10.2.2 Increase in Internet Participants
10.2.3 Risk Mitigation and IoT Safer
10.2.4 Increase in Risk
10.3 Future of the Internet of Things (IoT)
10.3.1 Internet of Things Market Size
10.3.2 The Future of IoT is AI
10.4 IoT Applications
10.4.1 IoT Solutions over the Next Five Years

Chapter-1: Introduction to IoT for Beginners

1.1 Introduction

The internet of things is a rapidly growing technology which aims connect all devices to the existing Internet infrastructure. At present only Mobiles, Computers, Tablets and Smart TV is connected with internet. By using IoT all the devices (Eg:- coffee maker, A.C, Washing Machine, Ceiling Fan, lights almost any thing you think of) having sensors can be connected with internet.

Internet (The world wide internet, used here in the form of a connection - could be through a WiFi or Mobile Data) + **Things** (Any electronic device that has the capability to sense)

Internet of Things: (Device(s) that have sensors, that are programmed to act in a certain way, are connected together to achieve a certain result.)

Uses: It is the network of items, which enables various objects to collect and exchange data with the help of internet.

Advantages: It will create an opportunity between physical world and computer based systems to result improved efficiency and benefits.

Illustration: Let's say your 1 year old suddenly starts crying, what do you do? you first start checking physically visible symptoms that may have hurt him/her OR make a guess that child is hungry and try feeding and if crying still doesn't stop than you may take him/her to the doctor. Basically since child cannot clearly communicate what exactly is happening to him/her you do some trial and errors (Of course over a period of time you understand the reason behind crying better). What if the child can tell even at age of 6 months that why I am upset or why I am crying or why I am not sleeping ? I will call it a smart child and that's what a smart object or thing would mean.

Things or objects around your who cannot speak but with help of sensors, some program and some processing of meaning attached to them, they can tell you what is happening with them.

1.2 Internet of Things (IoT)

The Internet of Things, or "IoT" for short, is about extending the power of the internet beyond computers and smartphones to a whole range of other things, processes, and environments.

"The Internet of Things (IoT) is a system of interrelated computing devices, mechanical and digital machines, objects, animals or people that are provided with unique identifiers and the ability to transfer data over a network without requiring human-to-human or human-to-computer interaction."

An Internet connection is something wonderful, it offers us all kinds of benefits that were not possible before. If you're old enough, think about your cell phone before it was a Smartphone. You can call and text, but now you can read any book, watch any movie, or listen to any song in the palm of your hand. And that's just to name a few of the amazing things your Smartphone can do. Connecting things to the internet produces many surprising benefits. We've all seen these benefits with our smartphones, laptops, and tablets, but this holds true for everything else, too. And yes, I am serious. The Internet of Things is actually a fairly simple concept; It means taking all the things in the world and connecting them to the internet.

1.2.1 Why IoT Matters

When something is connected to the Internet, that means you can send information or receive information, or both. This ability to send and / or receive information makes smart and smart things good. Let's use smartphones again as an example. Right now you can listen to almost any song in the world, but it's not because your phone actually has all the songs in the world stored. It's because all the songs in the world are stored elsewhere, but your phone can send information (asking for that song) and then receive information (transmit that song on your phone). To be smart, a thing doesn't need to have super storage or a supercomputer inside. All you have to do is connect to a super storage or a supercomputer. Being connected is amazing. In the Internet of Things, all things that connect to the Internet can be classified into three categories:

- Things that collect information and then send it.
- Things that receive information and then act on it.
- Things that do both.

The Internet of Things (IoT) is a network of physical devices that are connected to the Internet and collect and share information on a wireless network. These connection devices work on sensors.

1.2.2 Examples of IOT

Light Bulb: A sensor-based app can turn a light bulb on and off and makes human life a breeze.

Smart Home: In this case, all household appliances work through a sensor and provide more security and use smart household appliances such as lights, security and environmental controls that make human life easier and more convenient.

Healthcare and Fitness: In this area, IOT brought big changes and made everything at your fingertips, a smart watch can track your heartbeats and steps, and there are many more sensor-based apps.

1.2.3 Advantages of IOT

Now connectivity goes beyond laptops and phones, IOT will connect to cars, homes, smart cities, and healthcare. By 2025, connected devices across all technologies will reach 1.0 trillion.

YEAR	NUMBER OF CONNECTED DEVICES
1990	0.3 million
1999	90.0 million
2010	5.0 billion
2013	9.0 billion
2025	1.0 trillion

- It improves security because all devices are connected to each other and also to central devices, so they can be easily secured.
- It minimizes human efforts since all devices are connected to each other and communicate that it does a lot of work for humans.
- It reduces human efforts, then it also saves time and time is the most important factor that can be saved through IOT.
- IOT could replace the humans who are in charge of monitoring and maintaining supplies.
- Machine-to-machine interaction provides better efficiency, therefore accurate results can be obtained quickly.
- Information gathering is automated, so there is no longer a need to rely on human information. This results in valuable time savings.
- IOT technology continually collects information for correct decision making.

1.2.4 Disadvantages

- With this technology the violation of privacy increases. All data must be encrypted so that your personal data is not shared by everyone.
- IOT Advantage: It could replace the humans who monitor and maintain supplies. It also implies that people will lose jobs.
- We all now rely heavily on the Internet and technology in general. And with the increasing popularity of the Internet of Things, this dependency will also increase enormously. By relying more on the Internet, we could trigger a potentially catastrophic event if it fails.

1.3 Integration of Mobile Apps with IOT

Home automation is the best example of integrating mobile applications with IOT. In home automation, all household appliances are connected to a smartphone and a human being can easily operate important systems like lights, security and kitchen. Global smart home market is expected to grow by $ 53.45 billion by 2022.

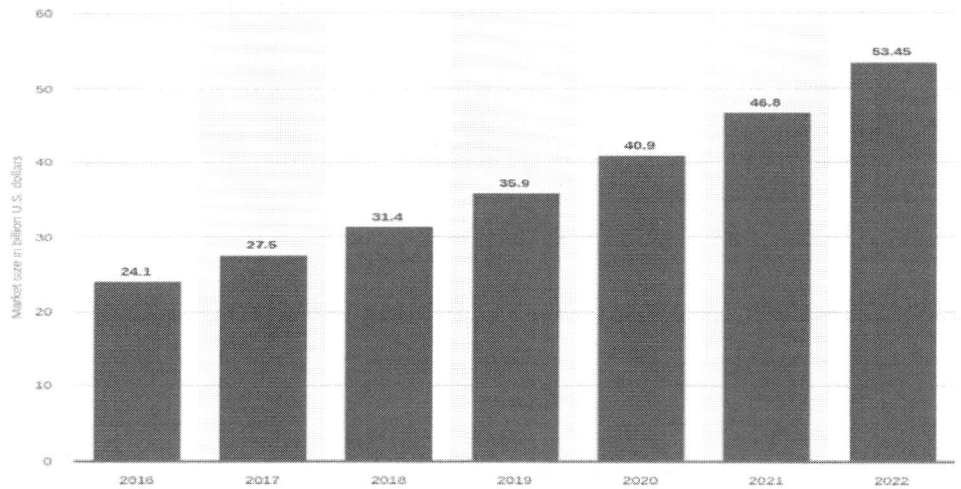

Figure 1.1: Global Smart Home Market

The Internet of Things is making human life easier with some smart apps that are very secure and work fast. With the help of IOT, mobile apps are advancing, such as chatbots, the use of artificial intelligence, and all facilities are working at the tip of our human fingers. The role of mobile apps in IOT makes all things fully automated. Today computers and the internet depend on humans for information. But this will change once the Internet of Things (IOT) enters the field with humans. IOT is a futuristic technology that claims to become a system of interrelated computing devices. It is a network of physical objects integrated with Internet connectivity that allows these objects to collect and exchange information.

The Thing's Internet name was first coined by Peter T. Lewis in a 1985 speech delivered to the US Federal Communications Commission (FCC). He states that "The Internet of Things or IOT is the integration of people, processes, and technology with pluggable devices and sensors to enable remote monitoring, status, manipulation, and trend assessment of such devices."

1.3.1 IPv6

The IOT will need more IP addresses to provide its benefits, but IPv4 cannot provide that many IP addresses. But this problem was already solved by introducing the IPv6 concept in the IOT. IPv6 expands the address space to 340 undecillion and this address space allows the Internet to be extended to everything.

1.3.2 Applications

The following are some possible areas where we can harness the power of the Internet of Things (IoT).

1. Smart Houses: Smart houses are generally in fashion among these applications. The vision of a cunning home is to control household machines, including lights, water flow from taps, home safety and well-being. The home owner gains admission to monitor and filter these activities from their smart devices (cell phones, tablets, workstations). Imagine a circumstance where you neglected to kill a water faucet, you are already out of your home. You can simply kill from your phone. Digital Homes allows you to manage all your home devices from one place. Also, a surveillance camera can be used under the IoT office to prevent burglary and theft.

2. Healthcare Solutions: IoT in Healthcare has opened new opportunities for teachers and clinical patients. Innovation enables specialists to continually access persistent clinical information, store it in the cloud, and offer it to others. It also reduces waiting time, helps verify accessibility of equipment and equipment, and untangles the procedure to distinguish constant ailments and perform the correct activities to alleviate the danger.

3. Smart city: Smart City or "City of the Future" is an idea that organizes innovation as responsible for offering improvements in the urban framework so that urban approaches are increasingly effective, less expensive and better for living. The idea of "Smart City" addresses open organization and organization through the mechanization of administrations in an imaginative and bearable way. The divisions that have been creating smart urban communities incorporate taxpayer-supported organizations, transport and traffic executives, vitality, medicinal services, water, inventive urban horticulture, and waste management.

4. Connected Car: Connected Car innovation is a huge and comprehensive system of different sensors, radio cables, implanted programming, and advancements that help correspondence explore our amazing world. You have a duty to decide on choices with consistency, accuracy, and speed. It must also be solid. These prerequisites will be significantly progressive when people fully hand over haggling control to self-governed or robotic vehicles that are, as a rule, effectively tested on our interstate highways at this time.

5. Smart Farming: Smart Farming is a frequently ignored IoT application. In any case, in light of the fact that the amount of farming tasks is generally remote and the large number of domesticated animals that ranchers attempt, all of this can be verified by the Internet of Things and can also alter the way what ranchers work. Be that as it may, this thought has not yet come to far-reaching consideration. In any case, despite everything, it is still one of the IoT applications that should not be underestimated. Smart farming can possibly become a significant field of application specifically in the agrarian article that ships nations.

6. Portable Devices: You should have a reasonable idea of the portable devices that are part of IoT technology and we are sure you have a couple of items as well. Google's famous company Glass has pulled out, but that hasn't reduced the odds of what innovation brings to the table. From exercise belts to shiny watches, everything you have associated with the web is an IoT piece. Through sensors once again these gadgets transmit information to give you generally accurate information

7. Smart Retail: Retailers have started to receive IoT fixes and use installed IoT frameworks in various apps that enhance store activities, for example expanding purchases, decreasing theft, enhancing inventory management and improve the buyer's shopping experience. Through IoT, physical retailers can tackle online challengers more unequivocally. They can recoup their lost share of the overall industry and attract shoppers by following these lines, making it easier for them to buy more and set aside cash.

8. Intelligent Environment: IOT will allow the detection of natural calamities early, such as the early detection of earthquakes through distributed control in specific places of tremors. We can also monitor CO_2 emissions from factories, pollution from cars, and toxic gases generated on farms. Also monitor the variation of the water level in rivers, dams and reservoirs.

1.4 Working of IoT

In simple terms, the Internet of Things is the name given to the entire spectrum of devices that are connected to the Internet, and / or each other and that can be turned on or off. These devices include obvious things like cell phones, computers, portable fitness devices, smart watches, and tablets; But it can also include devices like lamps and lighting fixtures, washing machines, refrigerators, automobiles, door locks, and even furniture, to name a few. Actually, these devices could be just about anything you can think of. The idea of IoT (Internet of Things) is to allow devices to connect, respond, adapt and influence each other, in order to make our lives easier and more convenient. Imagine, instead of sitting and waiting for your car to warm up on a freezing morning, your alarm clock "tells" your car that you are awake, which starts the heater to start heating it up. Or envision sensors on your furniture or floors that connect to the Internet and alert your lighting when it is on or off. All of these things are uses for IoT. High-speed broadband services are now widely accessible and the costs associated with them have decreased significantly in the past decade, meaning that companies are beginning to introduce Wi-Fi capabilities and sensors into devices and products at the stage manufacturing, knowing that at a time expanding market, could provide a great point of sale for potential customers.

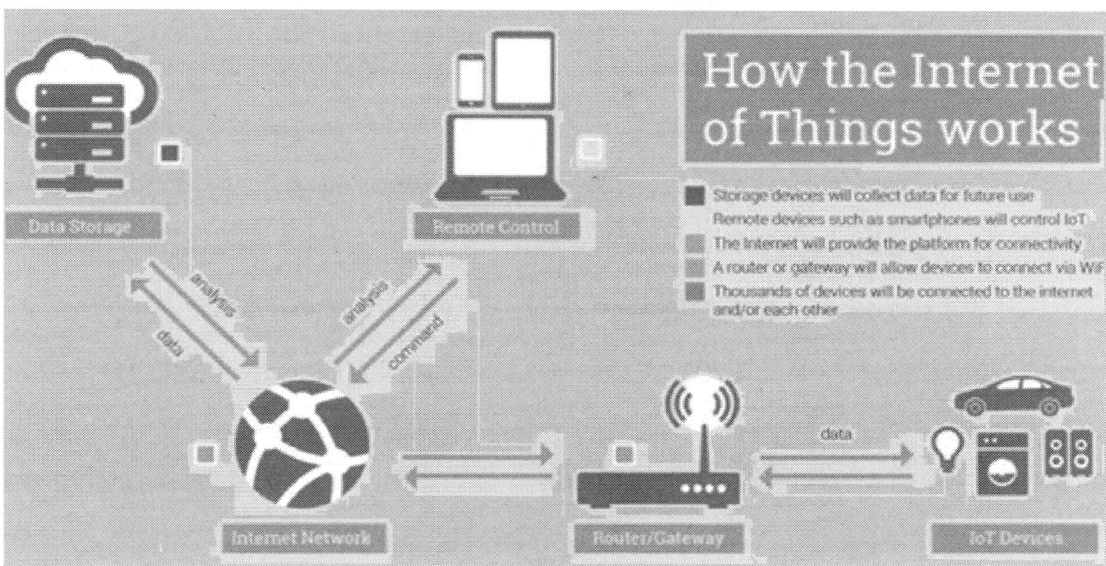

Figure 1.2: Block Diagram of IoT

Remote controls, such as smartphones, tablets, computers, and home control panels, connect to the Internet via Wi-Fi, which in turn is connected to each of your IoT-enabled devices. Data and analytics flow from the IoT device to the IoT device and back to the remote control, and can also be stored via the cloud, a local database, on the remote device, or on the IoT device.

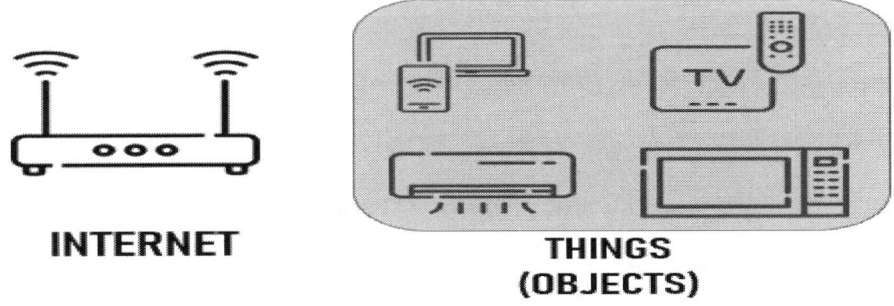

Figure 1.3: Working of IoT

Connectivity for the Internet of Things is established locally through the use of Ethernet, Bluetooth, Wi-Fi and, regarding the global connection, we can make use of an IP network. One of the simplest and best examples where IOT can be employed is at Smart Homes. We generally refer to each connected device as a smart object. All smart objects in a house can be controlled by a single device, which is usually a phone or tablet. In the following representation, there are several smart objects that are used in our homes and are connected to the Internet. A phone can control all the remaining devices. A washing machine can be turned on and off, the refrigerator temperature can be set, television channels can be changed, and many more functions can be controlled. Therefore, the potential to control all devices connected via the Internet at our fingertips has been made possible by the Internet of Things. We can make it more efficient by integrating technology with Artificial Intelligence.

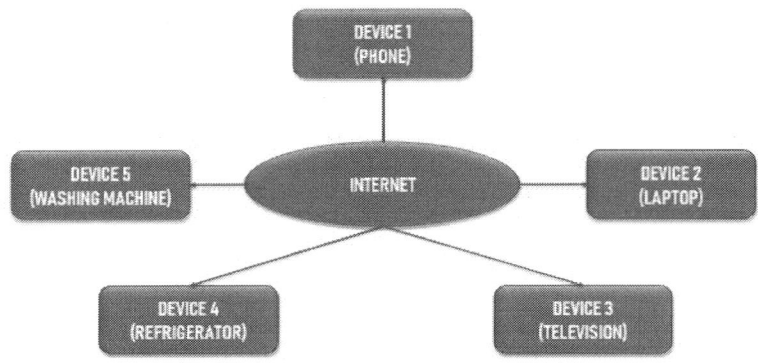

Figure 1.3: Architectuire of IoT

Therefore, the Internet of Things has established its trend and is growing at an accelerated rate in the technological world, and we will soon witness ultra-modern innovations in this field. IoT, also referred to as the Internet of Everything (IoE), is an environment of interrelated PC devices, computerized machines, and items that can move information progressively with each other, with the least human mediation. IoT solutions are made up of sensors / devices that "talk" to the cloud through availability. When the information reaches the cloud, scheduling it, and then you can choose to perform an activity, for example, send a warning, or consequently alter the sensors / devices without the customer's requirement. Be that as it may, if customer input is required or if the customer only needs to monitor the framework, a UI allows them to do just that. Any alteration or activity performed by the client is sent in another way through the framework: from the user interface, to the cloud and back to the sensors / devices to make a change. At the time we declare that the information is saved in the IoT stages, it does not imply that all the information is valuable. Gadgets carefully select only specific information that is applicable to run an activity. These pieces of data can distinguish examples, suggestions, and problems before they happen.

1.4.1 Challenges with the Internet of Things

Three main issues have arisen since manufacturers began developing IoT-enabled devices: security, privacy, and technology deployment.

Security: By allowing your remote devices to connect and manage so many of our valuable items, such as our cars, stereo systems, and kitchen appliances, and by being in control of secure devices like door and window locks, they must be connected to the Internet. It is vulnerable to hacking and unwanted interference. Not only that, but making connections via WiFi is one of the least secure ways to connect to any device. This puts our personal safety at risk. We may allow hackers to access our most valuable items, as well as data that explains when and where we use them, and details about other products that are connected to them. This could leave our valuable possessions vulnerable to thieves. Not to mention our own personal safety in case someone gains access to their home through an IoT device.

Privacy: This is another big problem with IoT. To perform basic functions and work effectively, devices must collect, analyze, and store our data. This is a big problem because online and even offline data storage is never completely secure. The sole purpose of many IoT devices will be to capture as much information about us as possible, inform our decisions, and simplify tasks. This poses a great risk to personal privacy and could make our data vulnerable to unwanted access.

Cost: For consumers to fully embrace the benefits of the Internet of Things, they will need to invest significantly in IoT-enabled devices, and while manufacturers may be starting to add technology to products as standard, it will inevitably be used as a marketing point to validate the increase in the cost of certain products. Not only that, but to have a fully connected home, people will need to replace a large number of non-IoT ready devices in their home. Easy perhaps for a first time buyer who is dedicated to creating a 'smart-home', but for a greater proportion of the population this would require a large budget to replace and revamp existing devices in their homes.

Chapter-2: Building Blocks of Internet of Things (IoT) and Their Characteristics

2.1 Trends and Characteristics

The main significant IoT trend in recent years is the explosive growth of connected and Internet controlled devices. The wide range of applications for IoT technology means that the details can be very different from one device to another, but there are basic characteristics shared by most. The IoT creates opportunities for a more direct integration of the physical world into computer-based systems, resulting in efficiency improvements, economic benefits, and reduced human efforts. The number of IoT devices increased 31% year-over-year to 8.4 billion in 2017 and it is estimated that there will be 30 billion devices by 2020. The global market value of IoT is projected to reach $ 7.1. trillion by 2020.

2.1.1 Architecture

The architecture of the IoT system, in its simplistic vision, consists of three levels: Level 1: devices, Level 2: Edge Gateway and Level 3: the cloud. Devices include network elements, such as sensors and actuators found in IoT equipment, particularly those that use protocols such as Modbus, Bluetooth, Zigbee, or proprietary protocols, to connect to an Edge Gateway. Edge Gateway consists of sensor data aggregation systems called Edge Gateways that provide functionality such as data preprocessing, security of cloud connectivity, use of systems such as WebSockets, the event center, and even some cases, edge analysis or fog computing. The Edge Gateway layer is also required to give a common view of the devices to the upper layers for easy administration.

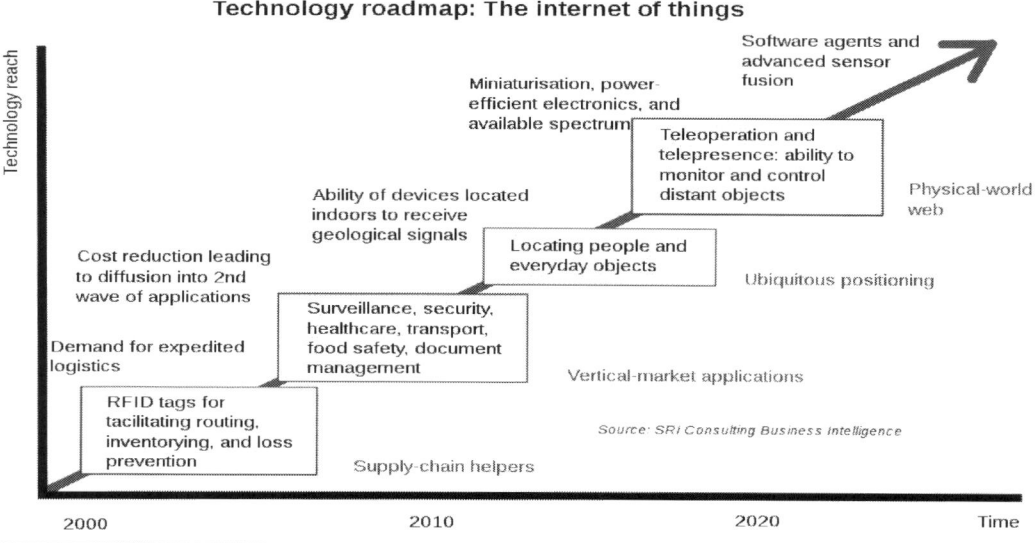

Figure 2.1: Technology Roadmap of IoT

The final tier includes the cloud application created for IoT using microservices architecture, which are generally polyglot and inherently secure in nature using HTTPS / OAuth. It includes several database systems that store sensor data, such as time series databases or asset stores that use back-end data storage systems (for example, Cassandra, PostgreSQL). The cloud level in most cloud-based IoT systems features a message and event queuing system that handles communication that is broadcast at all levels. Some experts classified the three levels in the IoT system as edge, platform and company, and these are connected by a proximity network, an access network and a service network, respectively. Building on the Internet of Things, the Network of Things is an Internet of Things application layer architecture that seeks to converge data from IoT devices into web applications to create innovative use cases. To schedule and control the flow of information on the Internet of Things, a planned architectural address is called BPM Everywhere, which is a combination of traditional process management with process mining and special capabilities to automate the control of large numbers of coordinated devices.

2.1.2 Network Architecture

The Internet of Things requires great scalability in the network space to handle the wave of devices. IETF 6LoWPAN would be used to connect devices to IP networks. With billions of devices added to the Internet space, IPv6 will play an important role in managing the scalability of the network layer. The IETF, ZeroMQ and MQTT restricted application protocol would provide light data transport. Fog computing is a viable alternative to avoid a large flow of data over the Internet. The computing power of edge devices to analyze and process data is extremely limited. Limited processing power is a key attribute of IoT devices as its purpose is to provide data on physical objects while remaining autonomous. Heavy processing requirements use more battery power, impairing the IoT's ability to operate. Scalability is easy because IoT devices simply supply data over the Internet to a server with sufficient processing power.

2.1.3 Application Layer

- **Near Field Communication (NFC):** Communication protocols that allow two electronic devices to communicate within a 4 cm range.
- **Radio Frequency Identification (RFID):** technology that uses electromagnetic fields to read data stored on labels embedded in other elements.
- **Wi-Fi:** Technology for local area networks based on the IEEE 802.11 standard, where devices can communicate through a shared access point or directly between individual devices.

- **ZigBee:** communication protocols for personal area networks based on the IEEE 802.15.4 standard, which provide low power consumption, low data rate, low cost and high performance.
- **Z-Wave:** wireless communication protocol used mainly for home automation and security applications

ADRC defines an application layer protocol and a support framework for implementing IoT applications.

1. **Short-range wireless**
   - **Bluetooth Mesh Networks** - A specification that provides a low-power Bluetooth Mesh Network (BLE) variant with a higher number of nodes and a standardized application layer (Models).
   - **Light-Fidelity (Li-Fi):** Wireless communication technology similar to the Wi-Fi standard, but which uses visible light communication to increase bandwidth.

2. **Medium-range Wireless**
   - **LTE-Advanced:** high-speed communication specification for mobile networks. It provides improvements to the LTE standard with extended coverage, higher performance and lower latency.
   - **5G:** 5G wireless networks can be used to meet high IoT communication requirements and connect large numbers of IoT devices, even when on the go.

3. **Long-range Wireless**
   - **Low Power Wide Area Networks (LPWAN):** Wireless networks designed to allow long-range communication at a low data rate, reducing the power and cost of transmission. Available LPWAN technologies and protocols: LoRaWan, Sigfox, NB-IoT, Weightless, RPMA.
   - **Very Small Aperture Terminal (VSAT):** Satellite communication technology that uses small-dish antennas for narrowband and broadband data.

4. **Wired**
   - **Ethernet :** A general-purpose network standard that uses twisted pair and fiber optic links in conjunction with hubs or switches.
   - **Power Line Communication (PLC):** Communication technology that uses electrical wiring to transport energy and data. Specifications like HomePlug or G.hn use PLCs for networking IoT devices.

2.2 IoT Architecture

The IoT architecture varies from one solution to another, depending on the type of solution we intend to build. IoT as technology consists mainly of four main components, on which an architecture is framed.

- Sensors
- Devices
- Door
- Cloud

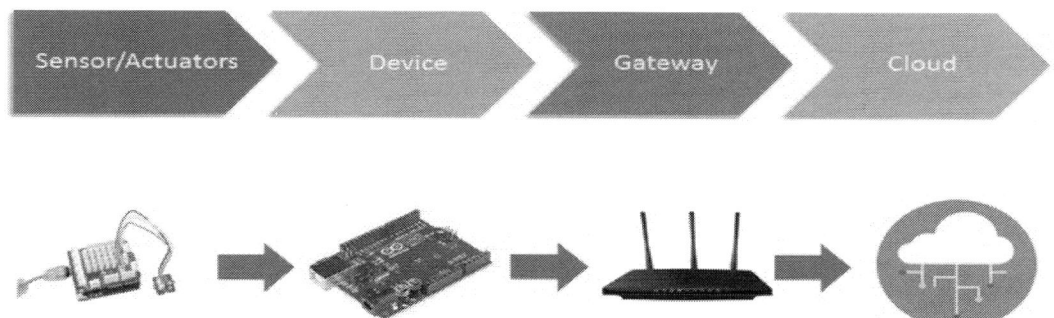

Figure 2.2: IoT Architecture

The following is the basic 4 Stage Architecture of IoT example:

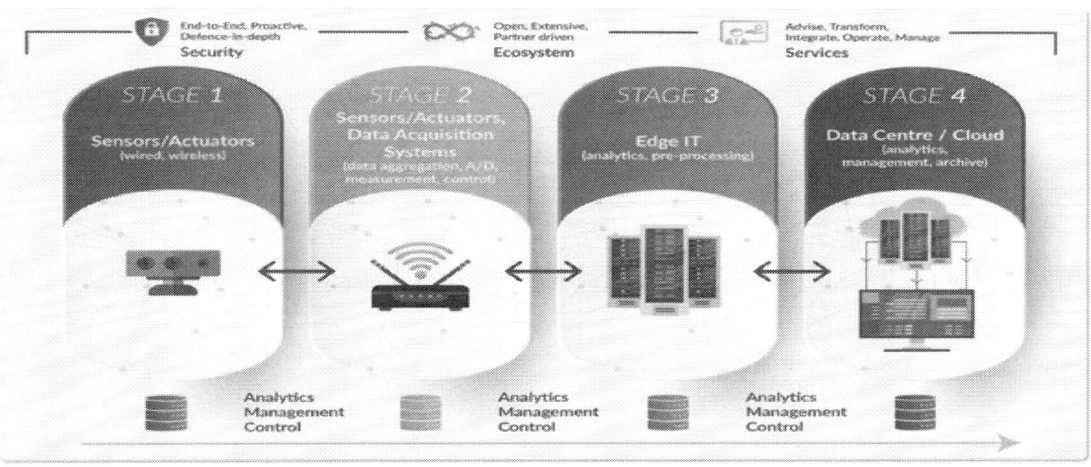

Figure 2.3: Basic 4 Stage Architecture of IoT

Stage 1:

Sensors: Sensors collect data from the environment or object under measurement and convert it to useful data. For example: gyroscope on mobile

Actuators: Actuators can also intervene to change the physical conditions that generate the data. An actuator could, for example, shut off a power source, adjust an air flow valve, or move a robotic clamp in an assembly process.

Detection / Performance: the stage covers everything. Example: Industrial devices for robotic camera systems, water level detectors, air quality sensors, accelerometers, and heart rate monitors

Stage 2:

systems are often found near sensors and actuators. For example: A pump can contain half a dozen sensors and actuators that feed data into a data aggregation device that also digitizes the data. This device can be physically connected to the pump. An adjacent gateway device or server will process the data and forward it to the Stage 3 or Stage 4 systems

Stage 3:

- Once the IoT data has been digitized and aggregated, it's ready to cross over to the IT realm
- However, the data may require additional processing before entering the data center.
- This is where state-of-the-art IT systems come in. They do more analysis.
- Edge IT processing systems can be located in remote offices or other edge locations, but are generally found in the facility or location where the sensors reside closest to the sensors, such as in a wiring closet

Stage 4:

- Stage 3 data is sent to the physical data center or cloud-based systems, where the most powerful IT systems can securely analyze, manage, and store the data.
- It takes longer to get results when you wait until the data reaches Stage 4, but you can run deeper analysis as well as combine your sensor data with data from other sources to get deeper insights.
- Stage 4 processing can be done on-premises, in the cloud, or in a hybrid cloud system, but the type of processing performed in this stage remains the same, regardless of platform

2.2.1 IoT Architecture Technology

Sensors : Sensors collect data from the environment or object under measurement and convert it to useful data. For example: gyroscope on mobile

Devices: Actuators can also intervene to change the physical conditions that generate the data. An actuator could, for example, shut off a power source, adjust an air flow valve, or move a robotic clamp in an assembly process.

Gateway: The data coming from the sensors must be prepared before they enter the final processing stage. Basically, data that is received in analogue form needs to be aggregated and converted to digital form, and this layer does exactly that with the help of an internet gateway that routes it through WLAN or other networks for further processing.

Cloud: It is to ensure the transition of data between field gateways and central IoT servers..

2.2.2 Working of IoT

The Internet of Things (IoT) consists of all web-enabled devices that collect, send, and act on the data they acquire from their surrounding environments using integrated communication hardware, processors, and sensors. These "connected" or "smart" devices can sometimes talk to other related devices and act on the information they get from each other. The Internet of Things is an environment of small smart hosting devices (because they connect to any device and are made by smart devices) that are always, anywhere, and anytime (IoT 3A) connected to each other and sending some data or Information that can be processed over the cloud to generate meaningful analytical results that can go a long way or trigger automatic action based on analysis. These little devices are called the "THING" of the Internet of Things and this environment consists of 3 ingredients called Device, Network and Application, also known as the Internet of Things DNA.

Internet "thing" of anything must qualify following criteria

- You should send some sensory data like pressure temperature humidity
- You must have a unique ID for you to identify yourself while communicating
- You should communicate with it and with the Internet gateway, as well as with WiFi

Internet of Things

1. Now some devices are already smart like your smartphone and the sensors already reside inside them, but for others we need to put some sensors

2. IPV6 or 6LoWPAN is already providing unique identification to these devices

3. For communication we need to add some IoT gateway as the most popular of many Texas Instrumentation CC3200s.

4. Next front-end as a mobile application or website where all data with analytical results or control device user interface is available.

Internet of things, Internet of everything (by CISCO), Smart Things (by IBM) are the terminologies coined by different companies for the same Internet of things. however, there are also some perceptions that the network of similar smart devices is IoT (just like the smartphone network), while a collected device network (like smart TV, smartphone, etc.) is Internet of everything. Internet of things or IoT (commonly known), elaborates the concept of the internet. It is the connection of various electronic devices that are equipped with sensors to transmit information or data to the device operator or other connected devices using the Global Telecommunications Unions Global Infrastructure Standards Initiative. This helps remotely detect and access devices for a detailed direct connection between computer-based systems and elements on the ground. This system eradicates the distance and obstacles involved in obtaining valuable data.

Imagine a room, outfitted with numerous computer systems and IT experts sitting at your desk, doing their jobs. View scenes from various movies that involve some research. A man comes in, he is the boss. He asks one of the IT experts for some information in his deep voice. The expert quickly types something and then waits a few seconds before answering his boss' query. Well the query answered in seconds is due to IoT. Actually, what happens here is that the expert writes specific words, codes, or statements for the query? With the help of sensors, the input machine sends the query to the machine located elsewhere and connected to the sensor. And again, with the help of sensors, the machine located elsewhere sends the data or information related to the query to the input machine. IoT (Internet of Things) is taking over the world in an unprecedented way. It's about bi-directionally connecting things to the internet to create a smarter, smarter world by building a network of smart things. Here things can be anything, including your toys, devices, and even your people. All living or non-living is considered as an object \ thing in IoT.

Similarly, when the internet is used to connect devices, vehicles, appliances, etc., it is called the internet of things. It involves extending Internet connectivity beyond standard devices, such as computers, androids, to any range of non-Internet-enabled physical devices and everyday objects. Integrated with technology, these devices can communicate and interact over the Internet, and can be remotely monitored and controlled.

Iot works with its 4 fundamental components:

1. Sensors- Collect data from the surrounding environment. The data collected can range from simple temperature monitoring to complex video. A device can have multiple sensors. Examples of sensors are GPS, camera, microphone, temperature sensors, etc. It also converts some physical phenomena into an electrical impulse.

2. Connectivity: The collected data is then sent to the cloud via a means of transport. Therefore, the sensors are connected to the cloud through various communication media such as wifi, satellite networks, bluetooth, wide area networks, etc.

3. Data processing: the collected data reaches the cloud, the software performs the processing of the acquired data. This can range from something very simple, like checking temperature readings on devices to very complex things like identifying objects.

4. User interface: The information is then made available to end users. This can be accomplished by notifying them through alarms, text messages, emails. If two devices are connected, actuators are used.

In iot network, one thing can have sensors and actuators or both processors and actuators or both sensors and processors or everything individually. The Internet of Things is the connection of devices (except computers and smart phones) to the Internet. Kitchen appliances, air conditioners, cars, and even heart monitors can be connected by implementing sensors. This technology has made a statement in various sectors, including manufacturing, healthcare, retail, automobiles, etc., and companies around the world are exploring the potential opportunities hidden within IoT.

2.3 Functions of 4 Stage Architecture of IoT

Sensors / Devices: Sensors / devices are the main components to consider in IoT technology. The sensors perceive all the minute details of the environment and collect all the data that can be used later. The sensors can be fused together, or they can be part of a device that does much more than just detect. For example, your phone is a device that comprises multiple sensors like GPS, accelerometer, camera, etc., but the phone is much more than a sensor.

Connectivity: The collected data is transferred to the cloud infrastructure which is also known as IoT platforms. But the device will need a medium like Bluetooth, Wi-Fi, cellular networks, WAN, etc. to transfer the data. The effectiveness and security of data depend largely on the speed and availability of these media.

Data Processing: Once the data reaches the cloud platform, it is analyzed correctly so that further action can be taken. This analysis can be as simple as monitoring AC temperature or complex, as using computer vision in the video to identify objects. IoT applications are developed in such a way that they can process easily and quickly.

User Interface: The last step is when the user receives a notification in the IoT mobile applications. Therefore, users will learn that their command has been executed through systems. It may sound easy, but you need a robust IoT platform or system that can be manually adjusted. For example, when the temperature of a refrigerator is not cold enough to freeze ice cubes, users should be able to do so without the system triggering manually.

The Internet of Things refers to interrelated devices that can transfer data over a network without the need for computing devices and human interaction. The Internet of Things is directly and indirectly associated with daily lifestyle products around the world. Internet connectivity is one of the main requirements of the different applications, as it is expected to grow at a high rate in the next forecast period. Additionally, Internet connectivity requires more monitoring through applications and human interactions. In addition, the Internet of Things offers easier and more comfortable control of different electronic devices from one place.

2.4 Advanced Technologies, Future Demand and Growth Analysis - IoT

IoT is expected to transform the way we work, live and whatnot. Smart objects and their different applications, such as smart cities, smart homes and smart environments, are dynamic in the evolution of the universal market. Furthermore, the growing demand for smart portable electronic products in different applications is driving the growth of the market. The potential for cyber physical systems to improve productivity in the production process and supply chain is increasing demand from the industrial sector in the Internet of Things market. Furthermore, rapid technological advancement is increasing the growth of the global IoT market. Wi-Fi connectivity is growing rapidly and due to various government initiatives supporting the widespread use of WiFi and collaboration between various companies, the market is experiencing an increase and is forecast to grow considerably during the forecast period. Consequently, all drivers will contribute to the expansion of the global Internet of Things market in the coming years. On the other hand, IoT is found in applications like urban planning and environmental planning, it could create immense opportunities for the IoT market.

2.4.1 Internet of Things, by Technology

- Zigbee
- Bluetooth low energy
- Near a communication field
- Wifi
- RFID

2.4.2 Need of IOT

Here is the answer for that:
1. Provides better scope for future data scientists
2. Provides promising career opportunities.
3. You also get real-time metrics, real data, and header connectivity across industries.

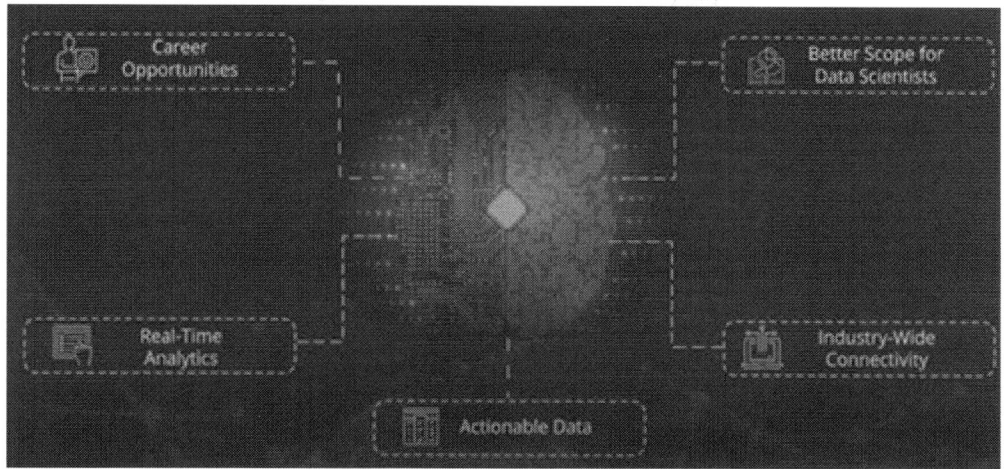

Figure 2.4: Need of IoT

2.4.3 Hardwares used in IOT

IOT devices hardware can be divided into the two categories

1. General devices -
a. They perform integrated processing and connectivity for the platforms they are connected to, either through a wired network or wireless interfaces.
b. They are the main component for data collection and information processing.
2. Sensing devices

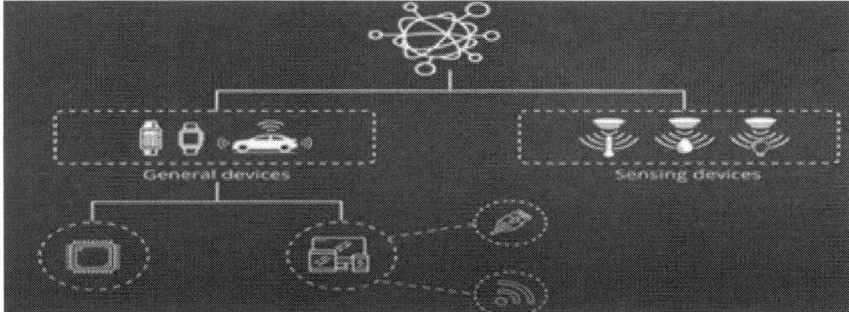

Figure 2.5: Hardware of IoT-Sensor

In addition to sensors, actuators are another important device at IOT that performs similar functions with different capabilities.

- Function as an interface between sensors and machines and collect various information such as humidity and light intensity
- This information is calculated using the edge layer that generally helps between the cloud and the sensor
- They are the layers that store the intermittent transfer of information.

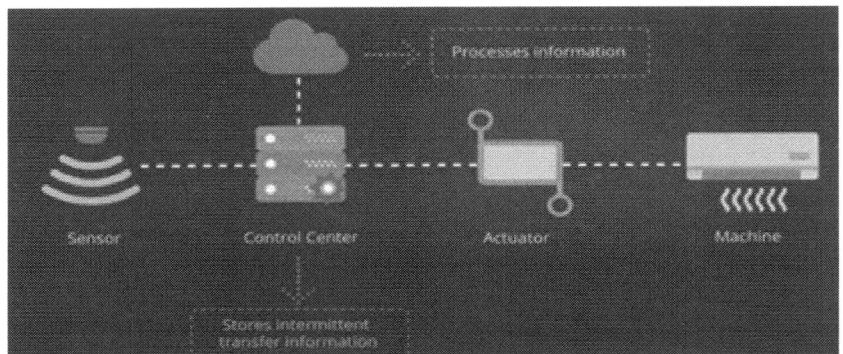

Figure 2.5: Hardware of IoT- Actuators

The second most crucial aspect of IOT is device management platforms or DMP

- DMPs are the platforms through which these assets interact with a software layer through network gateways.
- DMPs come with various functionalities including firmware updates, security patches, and metric reports.
- They also help develop alert mechanisms for industrial equipment with more open source OS like Arduino.

Figure 2.6: Hardware for IoT-DMP

Gateway Architecture

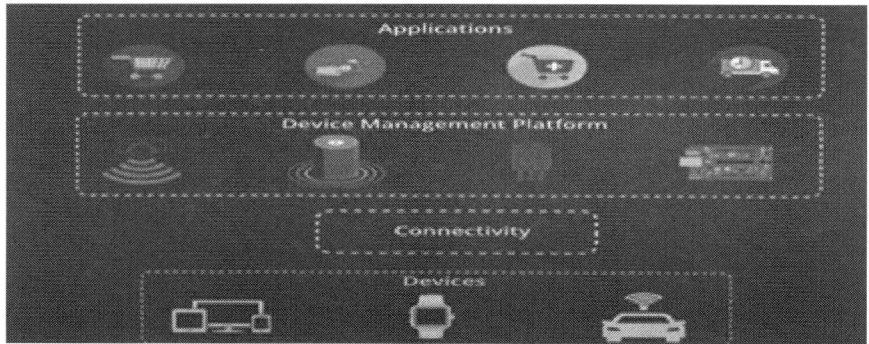

Figure 2.7: Gateway Architecture

Communication Protocols

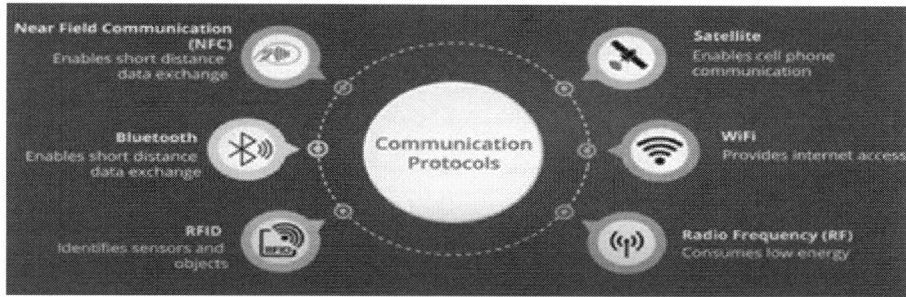

Figure 2.8: Communication Protocols

Future of IOT

The 5 most important areas where IoT will be used are-

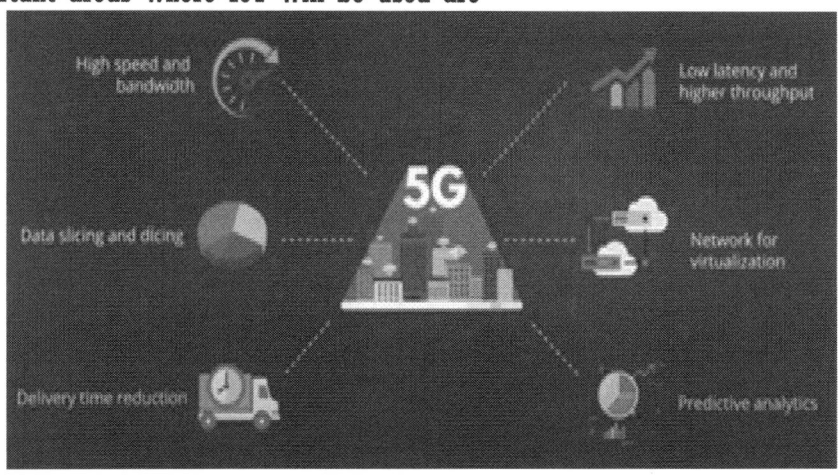

Figure 2.9: Future of IoT Connectivity

2.5 Design Steps involved in a Basic IoT System

Today, the Internet has become ubiquitous, has touched almost every corner of the world, and is affecting human life in unimaginable ways. However, the journey is far from over. We are now entering an era of even more widespread connectivity where a wide variety of devices will connect to the web. We are entering an era of the "Internet of Things" (abbreviated as IoT). The Internet of Things refers to a new type of world where almost all the devices and devices we use are connected to a network. We can use them collaboratively to accomplish complex tasks that require a high degree of intelligence. For this intelligence and interconnection, IoT devices are equipped with integrated sensors, actuators, processors, and transceivers. IoT is not a single technology; rather, it is an agglomeration of various technologies that work together. Sensors and actuators are devices that help interact with the physical environment. The data collected by the sensors must be stored and processed intelligently in order to deduce useful inferences from them. The Internet of Things finds various applications in healthcare, fitness, education, entertainment, social life, energy conservation, environmental monitoring, home automation, and transportation systems. In all these application areas, IoT technologies have been able to significantly reduce human effort and improve quality of life. The Internet of Things is also one of those smart advances that, in simple definition, means connectivity of objects to people through the exchange of data through the Internet, building a new infrastructure for the information society. In the Internet of things, all things can be classified into 2 categories.

1.Things that collect information and then send it: For example. The sensors can be temperature sensors, motion sensors, humidity sensors, air quality sensors, light sensors used in fields like

a. Retail: The IoT Indoor GPS system along with iBeacons and WiFi routers are helping retailers to better reach their customers and offer easy-to-use services.

b. Transportation: sensors are the new sensation and the companion of IoT technology. The transportation industry makes the most of it, from logistics to public transportation. Shipping vehicles can be tracked, monitored in real time, and geographic positioning data helps reduce traffic and accidents.

c. Agriculture: Sensor and actuator data helps farmers optimize their production and get the most out of it with less tendency to lose.

2. Things that receive information and then act on it. E.g. 3D printers widely used in the healthcare sector. IoT-enabled machines and devices in hospitals help doctors and staff monitor condition, improvement, and patient measurements more accurately and in a timely manner. In this life-changing era where technology is surpassing the human imagination, SMEs like Technostacks Infotech Pvt ltd also contribute with pride. They develop the IoT application that improves customer services, innovates new products, and transforms the world's Internet experience.

Imagine a world where all the other objects around us can speak, interact and work as a team. Technology has reached a level where each object, be it the car, the TV, the refrigerator, the geyser, the oven, etc., can work in sync with each other to provide the best comfort to humans and to carry the standard A new high. If you look, information technology has grown multiple times in the past two decades. With the advent of smartphones with internet access, technology has certainly set the stage for the aforementioned scenario, and has given a direction to growth, where not only will phones be smart, but any other object with which find will be smart. We can call an object "intelligent" if it can interact with humans and is access to the Internet. Not only can it perform its functions effectively, but also its performance and functions can be recorded and stored in a timely manner. For example. Take an example of a smart coffee machine. You can interact with humans through the touch screen to offer the best blend or variant of coffee, but since you are connected to the Internet, you can record user behavior, customer preferences and send maintenance alert messages to the owner.

The IoT definition says that "the Internet of Things is a concept or technology that aims to connect all devices to the Internet and help them communicate with each other using the Internet as a medium." The beauty of this technology is that its application is not linked to any particular industry. It can be expanded from a house to a housing company, from a machine to a manufacturing plant, from a city to the entire country. The essential elements of an IoT system are given below:

1) Sensors: If an object or machine can be smart, either by making it with a built-in sensor or you can get a sensor built into your existing machine or object. The basis of the integration of a sensor must be the quantity or the parameter that you want to verify, monitor, measure or record. The sensors are directly available or must be programmed to do what is necessary. The sensors may or may not be coupled to an actuator.

2) Network protocol or connectivity: once the sensors have registered the data, it must transmit it to a server or data warehouse. It can be done through wired mode (Ethernet) or wireless mode (Wi-Fi, MQTT protocol, Zigbee, CoAP, etc.). The mentioned wireless network protocols that are implemented depend on parameters such as data transmission speed, the volume of data to be transmitted, distance between the sensors, etc.

3) IoT Platform: An IoT platform is like an operating system that is installed on a computer or any electronic device to make it work and work. This is functional from the cloud and your responsibility is to control and interact with all connected devices, network protocols and manage the functionality of the entire IoT system. IoT platforms are offered as platforms as a service (Paas)

4) Infrastructure in the cloud: the data generated over a period of time could be in volume of Terabytes, Petabytes or even more, and having a company-owned server makes no technical and commercial sense, as it may run out of space very soon and also the maintenance cost of owning such a large server will be very high. The massive volume of data is stored on cloud servers and offered as Infrastructure as a Service (Iaas). All data is now stored in one format and is often linked to business software such as SAP or CRM, which can be accessed by the respective stakeholders, and this software is offered as a service such as Software as a Service (SaaS).

5) Results visualization - Collected and stored data is cleaned and formatted to be displayed as charts and pie charts, where you can draw the results and do business to make business decisions.

The Internet of Things (IoT) will transform the way industries work, their processes, work scenarios and research fields. As the number of smart objects or things around us increases, the speed of the Internet will increase proportionally and there will be an explosion of data that will need a massive cloud infrastructure for storage. The operations of each system will be interpreted as data and their results will draw inferences. Over time, this will be used to improve business decisions, increase operational efficiency, reduce downtime and outages. The steps include:

1. Firstly, data collection with the help of sensors (temperature, water level, etc.).
2. Send the data to the local processing unit (microcontrollers like Arduino)
3. Processing the data with the help of software like Arduino IDE.
4. Store the data in local storage.
5. Sending data to the cloud via internet (Ethernet, esp8266, NodeMCU).
6. Cloud processing followed by cloud storage.
7. Control your output devices from the servers.

Chapter-3: Domain specific IoT and Their Real-world Applications

3.1 Introduction

The Internet of Things is simply "A network of objects connected to the Internet capable of collecting and exchanging data." It is commonly abbreviated as IoT. The word "Internet of things" has two main parts; The Internet is the backbone of connectivity, and Things means objects / devices. In a simple way of saying it, it has "things" that perceive and collect data and send it to the Internet. This information may also be accessible by other "things". Let me give you a practical example. Let's imagine you have a "smart air conditioning unit" in your home that is connected to the Internet. (This is a "thing" connected to the Internet) Now, imagine it's a hot summer day and you've been home from work. You would like your home to be cool enough for when you enter. Therefore, when you leave your office, you can remotely turn on your home air conditioning unit using your mobile phone (another "Thing" connected to the Internet). Technically, with the Internet, you can control your AC system from anywhere in the world as long as both the AC and your mobile phone are connected via the "Internet". The Internet of Things (IoT) is the network of physical devices, vehicles, appliances and other elements integrated with electronics, software, sensors and connectivity that allows these objects to connect and exchange data. Each thing is uniquely identifiable through its integrated computer system, but can interact within the existing Internet infrastructure. The Internet of Things (IoT) is a network of devices and sensors connected to the Internet. These "things" (for example, your smart watch or smart thermostat) allow you to send and receive data without human intervention.

3.2.7 Connect with Things

With the implementation of the Internet of Things in the real world, there is a high possibility of a completely new way of interacting with and learning about things. Here in IoT everything will be connected to the sensors and will have a unique identity. These sensors will continuously send the data to the personal cloud to which it is connected.

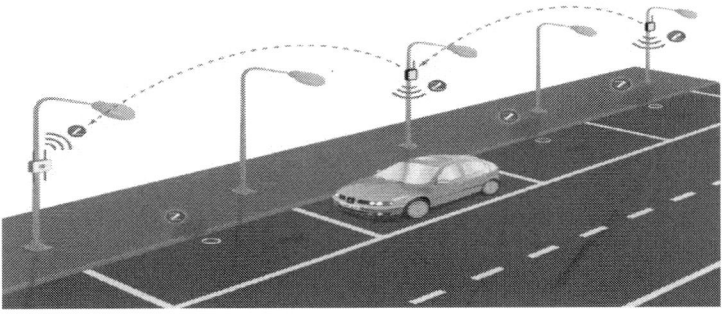

Figure 3.1: Example of IoT

3.1.2 Search for Things

Forget about Googling the information; you can even search for your lost things. Simply simple like "Where's my phone?"

3.1.3 Manage Things

According to surveys, 52% of the population lives in cities. With the implementation of IoT we can manage facilities for citizens very efficiently.

3.1.4 IoT stands Today

IoT is no longer a buzzword on the web, it has been making decent progress in verticals and it is happening in the real world. Geeks and technologists are already adapting and adopting the IoT in the form of portable devices.

Figure 3.2: IoT in the form of wearable devices

Google glass and Apple watch, Mi's Health band are the well-known in the field and to say as example.

3.3 Applications of IoT

IoT can change our lives by connecting various types of devices, not just computers, tablets, and communication devices to the Internet, it could lead to new ways of working with a wide range of systems, machinery, sensors, smart home devices, personal devices, and anything in between. it can be conected. The Internet of Things will play a crucial role in improving the world around us. Many technology giants like Apple Inc and Google are already employing technology in many unimaginable ways that will improve our lives. The impact of the Internet of things and technology in our lives is multiple and profound. Although the impact of technology was limited to labor-saving devices and devices that automatically turned on and off before, today's devices are much smarter than ever. They know their daily routine and use Artificial Intelligence and Machine Learning and can sense what they might need and suggest assistance regimes automatically. For example, automatic lawn mowers use sensors to determine the overall growth of the lawn on your lawn. This data is sent to motorized lawn mowers that cut the grass at the right time, ensuring optimal growth and allowing you to focus on garden design instead of spending half a day mowing the lawn. This data is then fed to motorized lawn mowers that cut the grass at the right time, ensuring optimal growth and letting you focus on the layout of the garden rather than spending half a day mowing it.

3.3.1 Smart Homes

This is one of the most discussed IoT applications. Basically, if you can connect all the devices in your home to the Internet, you can manage them remotely. For example, when you leave the office, turn on the car air conditioner, check everything you have in your fridge, open the curtains in your room, tell her to turn on the lights in 15 minutes and start recording the next episode of great theory of the explosion. All this while going down the stairs to your car. Companies have realized the potential of smart homes and have started working on them. Almost all electronic devices in homes will connect to your mobile or tablets to provide you with real-time data. Lighting accessories, thermostats, mirrors, toasters, automation systems, music systems, televisions, security, etc. They will connect to the mobile app and can allow you to control these devices from remote locations. In addition to providing real-time updates, these devices will be smart enough to perform tasks on their own over time based on user behavior. There certainly was a tremendous adoption of smart home technologies in 2016; Experts believe Amazon sold nine times as many Echos for the 2016 holiday season as it did the year before. Smart home technologies are expected to become even more important in 2017. Seventy percent of people who bought their first smart home device believe they are more likely to buy more, according to a Home Technology Survey. intelligent.

3.2.2 Home Automation and Smart Cities

Home automation and robotics, or Domotics, is a collection of technologies and systems that automate different tasks performed within a home to reduce human intervention as much as possible. This has given rise to the concept of smart homes. A smart home is part of the largest connected smart home ecosystem called Smart City. As the focus of city governance moves from areas to individual homes and even individuals, smart homes offer solutions like never before, ensuring safety, compliance, efficiency, adaptability and interactivity. For example, smart energy metering, combined with smart energy grids that power our homes, can regulate and monitor our energy use, taking advantage of unused additional energy that would otherwise be wasted, by powering smart cars that take us from point A to B. This shows how current devices are connected, smarter and more adaptable. As our devices learn our lifestyle patterns and reduce their dependence on us, task repetition decreases and efficiency increases. Everyday tasks like shopping for groceries and everyday items have been simplified. The IoT has the potential to transform entire cities by solving real problems that citizens face every day. With the right connections and data, the Internet of Things can solve traffic congestion problems and reduce noise, crime and pollution.

The idea of connecting cities and buildings with IoT devices has been one of the best so far, as it would solve problems related to lighting (streets and complexes), water collection, electricity, traffic, pollution levels, etc., with these smart devices being used only when necessary. According to the estimate, 75 million IoT devices will be shipped for IoT systems at 30% CAGR, from $ 36Bil in 2014 to $ 133bil in 2019. This investment will generate $ 421bil in economic value for cities around the world in 2019.

3.2.3 Environment Monitoring

Environmental conditions and anticipation of natural calamities can be improved. Wildlife parks can also be easily monitored by deploying drones and monitoring data in real time. This is already being done by WWF.

3.2.4 Medical and Healthcare

IoT devices can be used to enable remote health status monitoring and the emergency notification system. These health monitoring devices can range from blood pressure and heart rate monitors to advanced devices capable of monitoring specialized devices. Hospitals can monitor the health of their patients by collecting data from sensors connected to portable devices like the Health Band. From Apple to many other tech giants are already in this business. There are many health bands available in the market that can measure your health related data. IoT can improve the quality of patient care and health status monitoring. According to research, 646 million IoT devices will be deployed by 2020. These connected devices will collect data, automate processes, and more. The devices include portable devices, gym equipment and sensors at home, hospitals or remote locations through which real-time data will be sent to personal and medical users to better monitor the state of physical health.

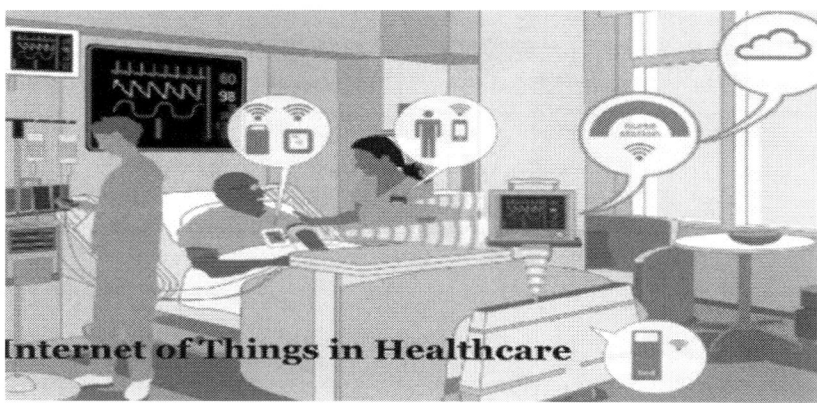

Figure 3.3: Internet of Things in Healthcare

Another field where IOT and automation have advanced is healthcare. With healthcare providers, insurance companies, and patients using the Internet much more than before, the cost of healthcare has been greatly reduced. As more and more patients use lifestyle and health management apps, patient data has become a breeze to gather as portable devices and innovative apps have come together to ensure patients add data relevant to your favorite devices or applications at your convenience. This ensures the accuracy of the data. This also motivates the user to keep fit and ensures that the portable device can track the data that the healthcare provider may need in emergencies. Many portable devices also offer message-triggering capabilities to emergency services.

3.2.5 Energy Management and Smart Grid

The power grids of the future will not only be smart enough but also highly reliable. The smart grid concept is becoming very popular worldwide. The basic idea behind smart grids is to collect data in an automated way and analyze the behavior of consumers and electricity providers to improve the efficiency and economy of electricity use. Smart grids that allow more homes and buildings to connect to the grid by solving many problems electricity departments face such as power theft, while also ensuring high-quality power and fewer blackouts.

3.2.6 Future of Banking

With portable banking applications, Omni channel customization, and the ability to detect fraud, all we can say is that IoT can redefine the future of banks through next-generation services.

3.2.7 Retail

The supply chain and retail will prove to be an important market in the use of IoT devices. From providing valuable information to retail store users about product fit, to how much product to buy for less waste is what IoT products are expected to do. With the launch of Amazon's Go on the market to forget the days when long lines take up most of our time and frustrate the shopping experience, we can expect the IoT to help both retailers and users. Knowing what the market demands are for retailers and what to buy and where for users will be easier. Products can be conveniently shipped to door steps based on weekly and monthly user behavior.

3.2.8 Transportation

With millions of vehicles on the ground, water and air, going online will be the next big innovation in the transportation industry, as this would reduce traffic, provide information on the location of stolen vehicles, and suggest an alternative route to avoid delays or unwanted through the data collected and analyzed. According to the research, there will be 220 million cars connected on the road by 2020.

3.2.9 Media and Advertising

Targeted advertising will take a whole new level with IoT by helping companies break free from ambiguous marketing activities. Based on your past behavior in certain categories of listening to news and seeing headlines in online media, you won't have to do anything when you get into your car to start traveling because your car's music system will automatically start reading the news of your choice to you. . While browsing the Internet, you will be shown relevant ads based on your needs or if an event of your choice is approaching.

3.2.10 Manufacturing, Oil, Gas and Mining

35% of manufacturers already use smart sensors and with the introduction of smart IoT devices the overall product production cycle for the market will be faster. This should increase competition as through updates and real-time data it will be easier to maintain the product in case of after sales services. In the oil and mining industry, it is expected that 5.4 million IoT devices will be installed by 2020 which will help in gathering environmental metrics for the extraction processes.

3.2.11 Construction

With the Introduction of IoT in construction industry, it will be easier and more efficient to track materials used in construction, manage stock and maintain optimum inventory levels. Elon Musk has already introduced smart roof tiles based on solar energy to generate more power and are aesthetically appealing as well.

3.2.12 Agriculture

IoT devices in agriculture will be installed in the soil to better manage and gather data in terms of acidity levels, temperature and variables that help in crop yields. It is estimated that 75million IoT

3.2.13 Fitness

Figure 3.4: Fitness

Fitness wearables are the new craze. With portable devices that offer levels of sophistication never before seen, users can monitor their progress toward a healthier lifestyle, form support groups, and stay motivated in the face of challenges. Walking, hydrating, and meditating reminders are increasingly used in workplaces where active employees have been shown to take fewer sick days off. Whether it's a style statement or just a new toy, wearables have helped people get up and find time for themselves in their busy and hectic lifestyles.

3.2.14 Security

Security devices have been around for a long time, but connected security devices have made our lives much easier. Smart cameras connected to the Internet can display data in real time, allowing us to be comfortable when we are not at home. Devices like nanny cameras now allow people to monitor their children while caregivers care for them when they are at home. Home security benefits a much of the IOT, as smarter devices can detect intrusions and automatically notify authorities.

Another subtle way to deter intruders is through the use of automated lighting and devices. Many devices can be configured to turn on at set times, leading potential intruders to think that occupants are home and deterring assault attempts. Today, it is difficult to imagine a world without connected devices. With bandwidth galore and the cost of wearable devices and smart devices shrinking from day to day, the use of these devices has increased. The positive impact of these devices has been an increased need for healthier lifestyles, fitter and better hydrated people who work to find time for themselves and their families.

Find unprotected spots

You must imagine and test all the risk situations that can happen to your system. Here is the model you can consider to verify:

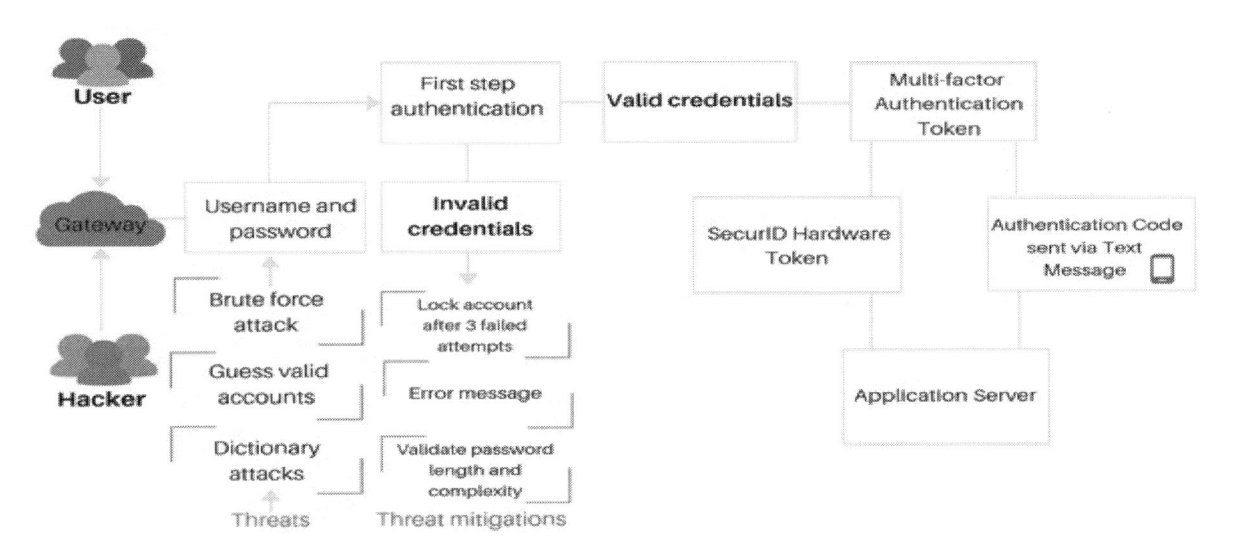

Figure 3.5: Authentication Method

Use a strong user Authentication Method
It has to be easy to use and safe at the same time. One of the best forms of security is to use a different password for each device.

Use OTA updates Timely

You should always update your system. There was an accident with a furnace that preheats to 400 degrees, all because the company skipped the planned system upgrade.

Limit the Network Access

You must control the remote use of your devices. Define how and who has access to the administration of your IoT system. It doesn't have to be easy to manage your system from any network or mobile phone.

Data Encryption

It is a mandatory way to protect your devices. You must encrypt the stored or transition data. The concept gained traction for its ability to connect the disconnected, the first physical objects that were previously unable to generate, transmit, and receive data unless augmented or manipulated. Embedding sensors, control systems, and processors in these objects enables horizontal communication through an open network of multiple nodes of first physical objects. The term is also loosely used to describe early digital connected devices, such as portable devices that can be classified as Digital Internet, while offering the same functionality as its physical counterpart developed in connected smart technology. The meaning and application of the term IoT will continue to evolve as new connected technologies emerge, replacing early physical objects with connected smart devices and use cases to make up all the new "Internet of X" classifications.

3.2.15 Wearable Technology

There were 78.1 million wearables sold in 2015 and the market is expected to grow to 411 million by 2020. All wearable technology, which includes smart watches, fitness trackers, VR headsets and more, generates a ton of data that businesses are just beginning to understand the possibilities and potential applications for.

3.2.16 Smart Cars

An astonishing 82 percent of cars are estimated to be connected to the Internet by 2021. Integration of apps, navigation and diagnostic tools, and even self-driving cars will be ways the Internet of Things is transforming the auto industry. The auto industry is investing heavily to determine the next IofT innovation.

3.2.17 Commercial Application

Business applications cover many industries, including medical and healthcare, V2X communications, transportation, and building automation. In the field of medical research, IoT is very essential as they elaborate on the process of having a vivid perspective. IoT is responsible for obtaining knowledge about emergency notification systems, remote health monitoring, medical diagnosis, etc.

3.2.18 Consumer Application

Devices sold under the consumer app generally connect vehicles, wearable technology, and connected health. IoT provides many savings, including long-term benefits. They automatically try to make sure the electronics and lighting are off. Another type of application for the consumer, apart from the smart home, is the care of the elderly, which includes help for people with disabilities and older people.

3.2.19 Infrastructure Application

The infrastructure application includes devices used in the fields of metropolitan-scale deployments, environmental monitoring, and energy management. IoT helps maintain the correct balance between cities and systems. Some of the devices that consume energy are: light bulbs, televisions, switches, etc. These devices work with Internet connectivity, which helps them maintain a balance between energy use and power generation. For the task of environmental monitoring, IoT devices tend to support environmental protection. One can quickly help monitor movements of wildlife and their habitats.

3.2.20 Industrial Application

The industrial sector is the fastest growing sector in the consumer and producer market. Manufacturing in industries addresses many IoT device requirements for ease of application. They can handle up to any digital control system. Digital control systems tend to perform some processes such as operator tools, service information controls, and process controls. IoT helps speed up the manufacturing of new products.

Chapter-4: Sensor and Actuator

4.1 Introduction

In recent years, sensor technology has been widely used in electronic measurement and signal processing instrumentation and research, and many other things. This rapid progress in Electronic Measurement and Instrumentation has created an increasing demand for personnel trained in Electronic Engineering. A sensor is a device that responds to any change in physical phenomena or environmental variables such as heat, pressure, humidity, movement, etc. This change affects the physical, chemical, or electromagnetic properties of the sensors, which are then processed in a more usable and readable form. . The sensor is the heart of a measurement system. It is the first element that comes into contact with environmental variables to generate an output.

Figure 4.1: Architecture of Sensor

4.1.1 Characteristics of Good Sensor

- High Sensitivity: Sensitivity indicates how much the device's output changes with the change of unit at the input (quantity to be measured). For example, the voltage of a temperature sensor changes by 1mV for each temperature change of 1oC that the sensitivity of the sensor is 1mV / oC.
- Linearity: the output must change linearly with the input.
- High Resolution: Resolution is the smallest change in input that the device can detect.
- Less noise and inconvenience.
- Less energy consumption.

A sensor is a hardware device that measures a physical quantity and produces a signal that can be read by an observer or an instrument. For example, a thermocouple converts the temperature into an output voltage, which can be read by a voltmeter. For greater precision, all sensors are calibrated to known standards. A transducer is a device that is powered by the power of one system and generally supplies power in another way to a second system. For example, a speaker is a transducer that transforms electrical signals into sound energy, or a thermocouple sensor transforms thermal energy into electrical energy. However, the words sensor and transducer are used synonymously, and application-specific names are assigned. In addition, suffixed derivatives ending in meters are used, such as accelerometer, flow meter and tachometer. Analog sensors produce a continuous output signal or voltage, which is generally proportional to the amount being measured. Physical quantities such as temperature, velocity, pressure, and displacement are all analog quantities, as they tend to be continuous in nature.

For example, the temperature of a liquid can be measured using a thermometer or thermocouple, which continually responds to changes in temperature as the liquid heats or cools. Transducers at the output of a system are often called actuators, and they convert the outputs of the electrical system into other types of energy, such as heat or mechanical energy. Digital sensors produce a discrete output signal or voltage that is a digital representation of the quantity being measured. Digital sensors produce a discrete (non-continuous) value, which can be output as a single bit or a combination of the bits to produce a single-byte output. A type of sensor related to the digital sensor is the smart sensor. Specifically, an integrated micro sensor with signal conditioning electronics such as analog-to-digital converters on a single silicon chip to form an integrated microelectromechanical component that can process information or communicate with a built-in microprocessor also known as a smart sensor.

Simple independent electronic circuits can be made to flash a light repeatedly or play a musical note. But for an electronic circuit or system to perform any useful task or function, it needs to be able to communicate with the "real world", either by reading an input signal from an "ON / OFF" switch or by activating some form of the output device to illuminate a single light. In other words, an electronic system or circuit must be able to "do" something and sensors and transducers are the perfect components to do it. The word "Transducer" is the collective term used for both sensors that can be used to detect a wide range of different forms of energy, such as motion, electrical signals, radiant energy, thermal or magnetic energy, etc., and actuators that can be use to change voltages or currents.

There are many different types of sensors and transducers, both analog and digital, and inputs and outputs available to choose from. The type of input or output transducer that is actually used depends on the type of signal or process that is "detected" or "controlled", but we can define a sensor and transducers as devices that convert one physical quantity into another. Devices that perform an "Input" function are commonly called Sensors because they "detect" a physical change in some characteristic that changes in response to a certain excitation, for example, heat or force, and convert it to an electrical signal. Devices that perform an "Output" function are generally called Actuators and are used to control some external device, for example, motion or sound. Electric transducers are used to convert energy of one type into energy of another type, for example, a microphone (input device) converts sound waves into electrical signals for the amplifier to amplify (a process), and a loudspeaker (output device) converts these electrical signals back to sound waves and an example of this type of simple input / output (I / O) system is shown below.

4.2 Input and Output System using Sound Transducers

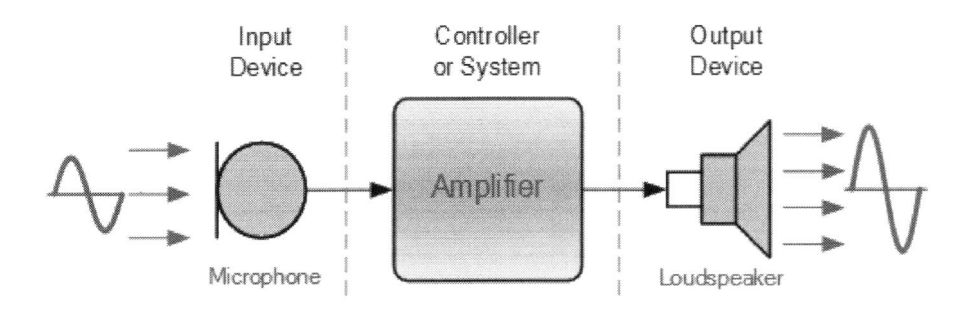

Figure 4.2: Input and Output System using Sound Transducers

4.2.1 Input Devices or Sensors

Sensors are "input" devices that convert a type of energy or quantity into an analog electrical signal. The most common forms of sensors are those that detect Position, Temperature, Light, Pressure and Speed. The simplest of all input devices is the switch or push button. Some sensors called "auto-generators" sensors generate output voltages or currents relative to the amount being measured, such as thermocouples and photovoltaic solar cells, and their output bandwidth is equal to the amount being measured. Some sensors called "modulating" sensors change their physical properties, such as inductance or resistance relative to the amount being measured, such as inductive sensors, LDRs, and potentiometers, and must be biased to provide an output voltage or current . Not all sensors produce a straight linear output and linearization circuits may be required. Signal conditioning may also be necessary to provide compatibility between the low-output signal from the sensors and the detection or amplification circuits. Some form of amplification is generally required to produce a suitable electrical signal that can be measured. Instrumentation type op amps are ideal for signal processing and conditioning the output signal from a sensor.

4.2.2 Output Devices or Actuators

"Output" devices are commonly called actuators, and the simplest of all actuators is the lamp. The relays provide good separation of low voltage electronic control signals and high power load circuits. The relays provide separation of the DC and AC circuits (i.e. switching an alternating current path via a DC control signal or vice versa). Solid state relays have a fast response, long life, no moving parts without contact arc or bounce, but require heat dissipation. Solenoids are electromagnetic devices that are primarily used to open or close pneumatic valves, safety gates, and robot-type applications. They are inductive loads, so a flywheel diode is required. DC motors with permanent magnets are cheaper and smaller than equivalent wound motors, as they have no field winding. Transistor switches can be used as simple single-pole ON / OFF controllers and pulse width speed control is obtained by varying the duty cycle of the control signal. Bi-directional motor control can be accomplished by connecting the motor within an H-bridge of the transistor. Stepper motors can be directly controlled using transistor switching techniques. The speed and position of a stepper motor can be precisely controlled using pulses so that it can operate in open loop mode. Microphones are input sound transducers that can detect acoustic waves, whether in Infra sound, audible sound, or the range of ultrasound generated by mechanical vibration. Speakers, buzzers, horns, and sirens are output devices and are used to produce an output sound, note, or alarm.

4.2.3 Types of Sensors

In almost all products for commercial and military applications, the number of sensors and transducers continues to increase. The application of "smart sensors" with digital communication techniques and the improvement of self-healing and improved precision sensor networks has created a growing application of sensor technology. Table 4.1 is a partial list of the applications for sensors and transducers.

Table 4.1: Typical Sensor Applications

Sensor Classification	Typical Sensor/Transducer
Thermal	Thermostat, Thermistor, thermocouple, thermopile
Force	Mechanical force, strain gauge, torque
Positional	Potentiometer, LVDT, rotary encoder
Fluid	Pressure, flow, viscometer
Optical	Photodiode, phototransistor, Photo detector, infrared, fiber optic
Motion	Displacement, velocity, acceleration, vibration, shock
Presence	Proximity
Environmental	Temperature, altitude, humidity, smoke

Input type transducers or sensors produce a voltage or signal output response that is proportional to the change in the amount that they are measuring (the stimulus). The type or amount of the output signal depends on the type of sensor being used. But in general, all types of sensors can be classified into two types, either passive or active sensors. There are many different types of sensors and transducers available on the market, and the choice of which one to actually use depends on the amount being measured or controlled.

Table 4.2: Common Sensors and Transducers

Quantity being	Input Device (Sensor)	Output Device
Light Level	Light Dependent Resistor (LDR)	Lights & Lamps
Temperature	Thermocouple	Heater
Force/Pressure	Strain Gauge	Lifts & Jacks
Position	Potentiometer	Motor
Speed	Tacho-generator	AC and DC Motors
Sound	Carbon Microphone	Bell

In general, active sensors require an external power supply to function, called an excitation signal that the sensor uses to produce the output signal. Active sensors are self-generated devices because their own properties change in response to an external effect that produces, for example, an output voltage of 1 to 10v DC or an output current such as 4 to 20 mA DC. Active sensors can also produce signal amplification. A good example of an active sensor is an LVDT sensor or a voltage meter. Voltage meters are networks of pressure sensitive resistive bridges that are externally polarized (drive signal) such that they produce an output voltage in proportion to the amount of force and / or voltage applied to the sensor. Unlike an active sensor, a passive sensor does not need any additional power source or drive voltage. Instead, a passive sensor generates an output signal in response to some external stimulus. For example, a thermocouple that generates its own voltage output when exposed to heat. So passive sensors are direct sensors that change their physical properties, such as resistance, capacitance or inductance, etc. But in addition to analog sensors, digital sensors produce a discrete output that represents a number or binary digit, such as a logical level "0" or a logical level "1".

4.2.4 Analogue and Digital Sensors

Analogue Sensors

Analog sensors produce a continuous output signal or voltage that is generally proportional to the amount being measured. Physical quantities such as temperature, speed, pressure, displacement, deformation, etc. they are all analog quantities as they tend to be continuous in nature. For example, the temperature of a liquid can be measured using a thermometer or thermocouple that continually responds to changes in temperature as the liquid heats or cools.

Thermocouple used to produce an Analogue Signal

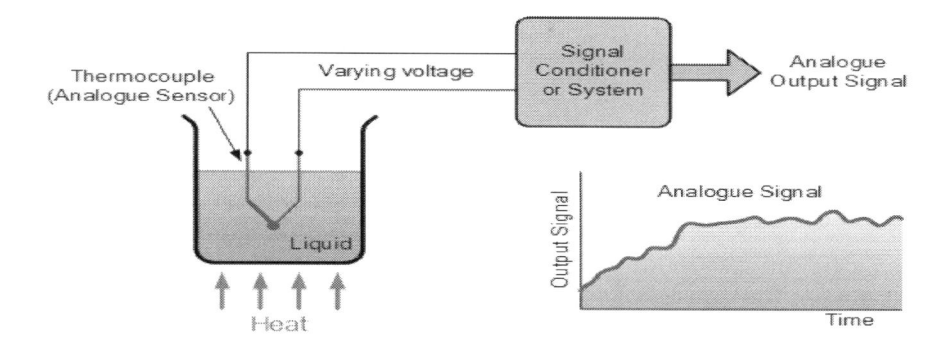

Figure 4.3: Thermocouple used to produce an Analogue Signal

Analog sensors tend to produce output signals that change smoothly and continuously over time. These signals tend to be very small in value, from a few microvolts (uV) to several millivolts (mV), so some form of amplification is required. So circuits that measure analog signals generally have slow response and / or low precision. Additionally, analog signals can be easily converted to digital type signals for use in microcontroller systems by using analog to digital converters or ADCs.

Digital Sensors

Digital sensors produce discrete digital output signals or voltages that are a digital representation of the quantity being measured. Digital sensors produce a binary output signal in the form of a logical "1" or a logical "0", ("ON" or "OFF"). This means that a digital signal only produces discrete (non-continuous) values that can be output as a single "bit" (serial transmission) or by combining the bits to produce a single "byte" output (parallel transmission).

Light Sensor used to produce an Digital Signal

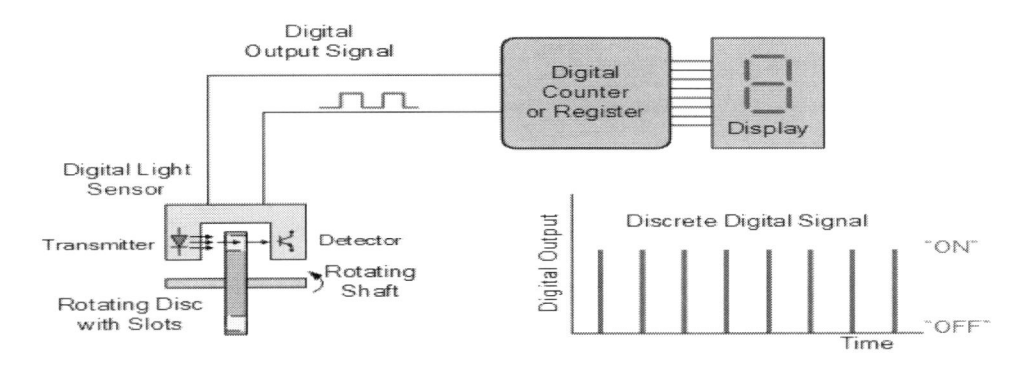

Figure 4.4: Light Sensor used to produce a Digital Signal

In Figure 4.4, the speed of the rotary axis is measured using a digital LED / optodetector sensor. The disk that is fixed to a rotating shaft (for example, from a motor or robot wheels), has a series of transparent grooves within its design. As the disk rotates with spindle speed, each slot passes through the sensor in turn producing an output pulse that represents a logical "1" or logical "0" level. These pulses are sent to a counter register and finally to an output display to show the speed or revolutions of the shaft. By increasing the number of slots or "windows" within the disc, more output pulses can be produced for each revolution of the shaft. The advantage of this is that higher resolution and precision are achieved as fractions of revolution can be detected. Then, this type of sensor arrangement could also be used for positional control with one of the disc slots representing a reference position.

Compared to analog signals, digital signals or quantities have very high precision and can be measured and "sampled" at a very high clock speed. The precision of the digital signal is proportional to the number of bits used to represent the measured quantity. For example, using an 8-bit processor will produce an accuracy of 0.390% (1 part in 256). While the use of a 16-bit processor provides an accuracy of 0.0015% (1 part in 65,536) or 260 times more accurate. This precision can be maintained as digital quantities are handled and processed very quickly, millions of times faster than analog signals. In most cases, sensors and, more specifically, analog sensors generally require an external power supply and some form of additional signal amplification or filtering to produce a suitable electrical signal that can be measured or used. A very good way to achieve both amplification and filtering within a single circuit is to use op amps as seen above.

4.2.5 Signal Conditioning of Sensors

Operational amplifiers can be used to provide signal amplification when connected in inverted or non-inverted configurations. Very small analog signal voltages produced by a sensor, such as a few millivolts or even picovolts, can be amplified many times by a simple op amp circuit to produce a much higher voltage signal of, say, 5v or 5mA which can then be used as an input signal to a microprocessor or analog-to-digital based system. Therefore, to provide any useful signal, the output signal from a sensor must be amplified with an amplifier that has a voltage gain of up to 10,000 and a current gain of up to 1,000,000 with signal amplification being linear with the signal. output as an exact reproduction of the input, just changed in amplitude. So amplification is part of signal conditioning. Therefore, when using analog sensors, some form of amplification (gain), impedance matching, isolation between input and output, or perhaps filtering (frequency selection) may generally be required before the signal can be used. , and this is conveniently done by op amps.

Furthermore, by measuring very small physical changes, the output signal from a sensor can be "contaminated" by unwanted signals or voltages that prevent the actual required signal from being measured correctly. These unwanted signals are called "noise". This noise or interference can be greatly reduced or even eliminated through the use of signal conditioning or filtering techniques. By using a low-pass or high-pass or even band-pass filter, the "bandwidth" of the noise can be reduced to leave only the required output signal. For example, many types of inputs for switches, keyboards, or manual controls are not capable of rapid state change, so a low-pass filter can be used. When the interference is on a particular frequency, for example, network frequency, narrowband rejection, or Notch filters can be used to produce frequency selective filters.

Typical Op-amp Filters

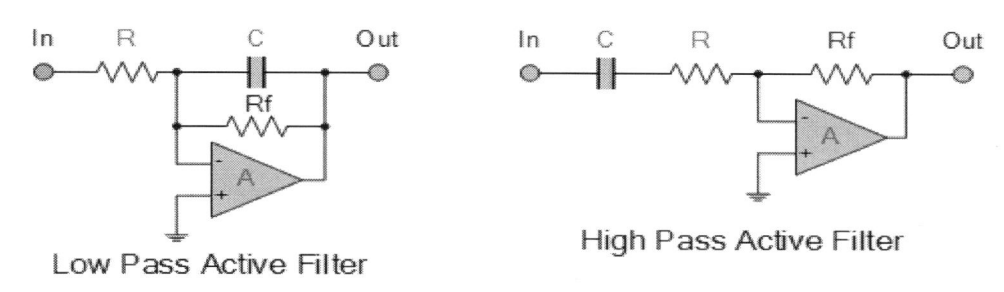

Figure 4.5 : Typical Op-amp Filters

If some random noise remains after filtering, it may be necessary to take several samples and then average them to obtain the final value, thus increasing the signal / noise ratio. Either way, both amplification and filtering play an important role in connecting sensors and transducers to microprocessor-based and electronic systems in "real world" conditions.

4.3 Capacitive Displacement Sensor

Capacitive displacement sensors "are non-contact devices capable of measuring the position and / or changing the position of any conductive target in high resolution." They can also measure the thickness or density of non-conductive materials. Capacitive displacement sensors are used in a wide variety of applications, including semiconductor processing, the assembly of precision equipment such as disk drives, precision thickness measurements, machine tool metrology, and assembly line testing. These types of sensors can be found in machining and manufacturing facilities around the world.

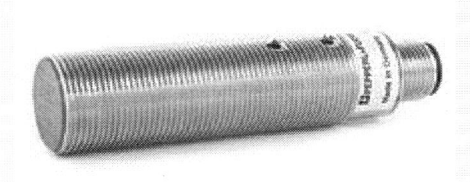

Figure 4.6: Industrial capacitive sensor

4.3.1 Basic Capacitive Theory

Capacitance is an electrical property that is created by applying an electrical charge to two conductive objects with a space between them. A simple demonstration is two parallel conductor plates of the same profile with a gap between them and a load applied to them. In this situation, capacitance can be expressed by the equation:

$$C = \frac{\epsilon_o KA}{d} \quad (4.1)$$

Where C is the capacitance, ε0 is the permittivity of the free space constant, K is the dielectric constant of the material in space, A is the area of the plates, and d is the distance between the plates. There are two general types of capacitive displacement detection systems. One type is used to measure thicknesses of conductive materials. The other type measures thickness of non-conductive materials or the level of a fluid.

A capacitive detection system for conductive materials uses a model similar to that described above, but instead of one of the conductive plates, it is the sensor, and instead of the other, it is the conductive target to be measured. Since the area of the probe and target remain constant, and the dielectric of the material in space (usually air) also remains constant, "any change in capacitance is the result of a change in the distance between the probe and target " Therefore, the above equation can be simplified to:

$$C \propto 1/d \quad (4.2)$$

Where α indicates a proportional relationship. Due to this proportional relationship, a capacitive detection system can measure changes in capacitance and translate these changes into distance measurements. The operation of the sensor to measure the thickness of non-conductive materials can be considered as two capacitors in series, each with a different dielectric (and a dielectric constant). The sum of the thicknesses of the two dielectric materials remains constant, but the thickness of each can vary. The thickness of the material to be measured displaces the other dielectric. Space is often an air space (dielectric constant = 1) and the material has a higher dielectric. As the material becomes thicker, the capacitance increases and is detected by the system. A sensor to measure fluid levels works like two condensers in parallel with a constant total area. Again, the difference in the dielectric constant of the fluid and the dielectric constant of the air results in detectable changes in capacitance between the conductive probes or plates. The capacitance C between the two capacitive transducer boards is given by:

C = $\epsilon_o \epsilon_r$ A/ d   (4.3)

Where,

C = capacitance of the capacitor or the variable capacitance transducer

$\varepsilon_0 =$ absolute permittivity
$\varepsilon_r =$ relative permittivity
$\varepsilon_0 \varepsilon_r =$ dielectric constant of the capacitive transducer
A = area of the plates
d = distance between the plates

From the above formula it follows that the capacitance of the capacitive transducer depends on the area of the plates and the distance between them. The capacitance of the capacitive transducer also changes with the dielectric constant of the dielectric material used in it. Therefore, the capacitance of the variable capacitance transducer can change with the change of the dielectric material, the change in the area of the plates and the distance between plates. Depending on the parameter you change for the capacitive transducers, they are of three types as mentioned below.

(1) Change of the type of dielectric constant of capacitive transducers: in this capacitive transducer, the dielectric material between the two plates changes, due to which the capacitance of the transducer also changes. When the input quantity to be measured changes, the value of the dielectric constant also changes, so the capacitance of the instrument changes. This capacitance, calibrated against the input quantity, directly provides the value of the quantity to be measured. This principle is used to measure the level in the hydrogen container, where the change in the hydrogen level between the two plates results in the change in the dielectric constant of the capacitance transducer. In addition to the level, this principle can also be used to measure the humidity and moisture content of the air.

(2) Changing area of capacitive transducer plates: The capacitance of the variable capacitance transducer also changes with the area of the two plates. This principle is used in the torque meter, used to measure the torque on the shaft. This comprises the sleeve having axially cut teeth and the corresponding axis having similar teeth on its periphery.

Differential capacitive transducers

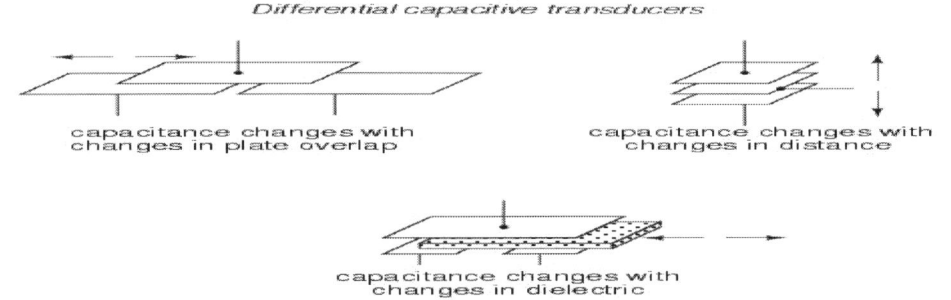

capacitance changes with changes in plate overlap

capacitance changes with changes in distance

capacitance changes with changes in dielectric

Figure 4.7: Differential Capacitive Transducer

(3) Change of distance between the plates of the capacitive transducers: in these capacitive transducers, the distance between the plates is variable, while the area of the plates and the dielectric constant remain constant. This is the most widely used type of variable capacitance transducer. To measure the displacement of the object, one plate of the capacitance transducer is held fixed, while the other is connected to the object. When the object moves, the capacitance transducer plate also moves, this results in a change in the distance between the two plates and a change in capacitance. The modified capacitance is easily measured and calibrated against the input quantity, which is the displacement. This principle can also be used to measure pressure, speed, acceleration, etc.

4.3.2 Advantages and Disadvantages of Capacitive Transducer

Advantages: Produces an accurate frequency response to both static and dynamic measurements.

Disadvantages: an increase or decrease in temperature at a high level will change the precision of the device. Since the cable is long, it can cause errors or distortion in the signals.

4.3.3 Applications of Capacitive Transducer

1. **Precision Positioning:** One of the most common applications of capacitive sensors is precision positioning. Capacitive displacement sensors can be used to measure the position of objects down to the nanometer level. This type of precise positioning is used in the semiconductor industry where silicon wafers must be placed for exposure. Capacitive sensors are also used to pre-focus electron microscopes used in wafer testing and examination.

2. **Disk Drive Industry:** In the disk drive industry, capacitive displacement sensors are used to measure the exhaustion (a measure of how much the axis of rotation deviates from an ideal fixed line) of the spindles of the disk drive. By knowing the exact exhaustion of these spindles, disk drive manufacturers can determine the maximum amount of data that can be placed on the drives. Capacitive sensors are also used to ensure that the disk drives are orthogonal to the axis before data is written to them.

3. **Precision thickness measurements:** Capacitive displacement sensors can be used to make very accurate thickness measurements. Capacitive displacement sensors operate by measuring position changes. If the position of a reference part of known thickness is measured, other parts can be subsequently measured and the position differences can be used to determine the thickness of these parts. For this to be effective using a single probe, the parts must be completely flat and measured on a perfectly flat surface. If the part to be measured has any curvature or deformity, or simply does not rest firmly against the flat surface, the distance between the part to be measured and the surface on which it is placed will be wrongly included in the thickness measurement. This error can be eliminated by using two capacitive sensors to measure a single part. Capacitive sensors are placed on both sides of the piece to be measured. When measuring parts from both sides, curvature and deformities are taken into account in the measurement and their effects are not included in thickness readings.

4. **Non-conductive objectives:** The thickness of plastic materials can be measured with the material placed between two electrodes at a set distance. These form a type of capacitor. The plastic when placed between the electrodes acts as a dielectric and displaces air (which has a dielectric constant of 1, different from plastic). Consequently, the capacitance between the electrodes changes. Capacitance changes can be measured and correlated with the thickness of the material. Capacitive sensor circuits can be built that are capable of detecting changes in capacitance on the order of 10 to 5 Farads Pico.

5. **Non-conductive lenses:** Although capacitive displacement sensors are most often used to detect changes in the position of conductive lenses, they can also be used to detect the thickness and / or density of non-conductive lenses. A non-conductive object placed between the probe and the conductive target will have a different dielectric constant than the air in space and will therefore change the capacitance between the probe and the target. (See the first equation above) By analyzing this change in capacitance, the thickness and density of the non-conductor can be determined.

6. **Machine tool metrology:** Capacitive displacement sensors are often used in metrology applications. In many cases, sensors are used "to measure shape errors in the part being produced. But they can also measure errors that arise in the equipment used to manufacture the part, a practice known as machine tool metrology." In many Cases are used to analyze and optimize spindle rotation on various machine tools, for example, surface grinders, lathes, milling machines, and air bearing spindles. By measuring errors on the machines themselves, rather than simply measure errors in final products, problems can be fixed and fixed earlier in the manufacturing process.

7. **Assembly Line Testing:** Capacitive displacement sensors are often used in assembly line testing. They are sometimes used to test assembled parts for uniformity, thickness, or other design characteristics. At other times, they are simply used to search for the presence or absence of a certain component, such as glue. Using capacitive sensors to test assembly line parts can help avoid quality problems later in the production process.

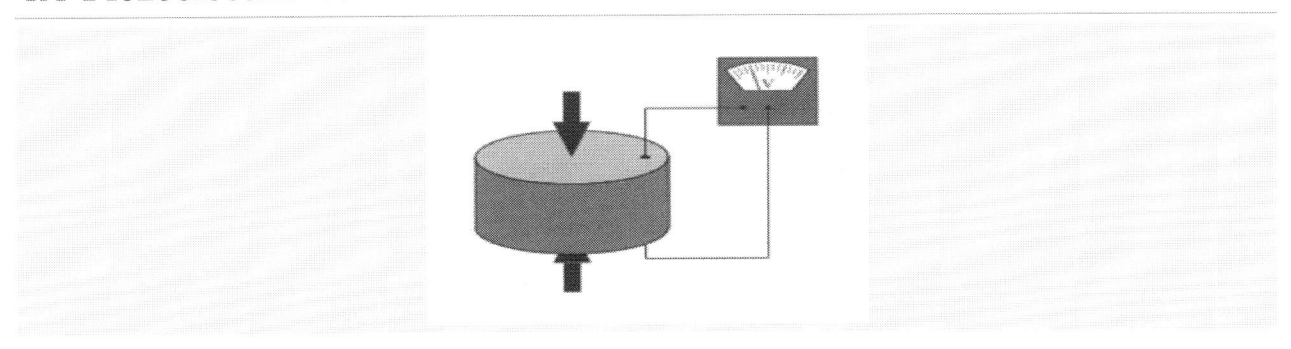

4.4 Piezoelectric Sensor

Figure 4.8: A Piezoelectric Sensor

A piezoelectric sensor is a device that uses the piezoelectric effect to measure changes in pressure, acceleration, temperature, tension, or force by converting them to an electrical charge. The prefix piezo is Greek for 'press' or 'squeeze'. Piezoelectric transducers operate on the principle of the piezoelectric effect. When mechanical forces or forces are applied to some materials along certain planes, they produce electrical voltage. This electrical voltage can be easily measured with voltage measuring instruments, which can be used to measure effort or force. Physical amounts like stress and strength cannot be measured directly. In such cases, material exhibiting piezoelectric transducers can be used. The tension or force to be measured is applied along certain planes to these materials. The voltage output obtained from these materials due to the piezoelectric effect is proportional to the applied voltage or force. The output voltage can be calibrated against the applied voltage or force so that the measured value of the output voltage directly provides the value of the applied voltage or force. In fact, the scale can be directly marked in terms of effort or force to give the values directly. The voltage output obtained from the materials due to the piezoelectric effect is very small and has a high impedance. Some amplifiers, an auxiliary circuit and connecting cables are required to measure the output.

4.4.1 Piezoelectric Effect

There are certain materials that generate electrical potential or voltage when mechanical stress is applied to them or, conversely, when voltage is applied to them, they tend to change dimensions along a certain plane. This effect is called as the piezoelectric effect. This effect was discovered in 1880 by Pierre and Jacques Curie. Some of the materials that exhibit a piezoelectric effect are quartz, Rochelle's salt, polarized barium titanate, ammonium dihydrogen, common sugar, etc.

4.4.2 Applications of Piezoelectric Sensor

Piezoelectric sensors are versatile tools for measuring various processes. They are used for quality assurance, process control, and for research and development in many industries. Pierre Curie discovered the piezoelectric effect in 1880, but only in the 1950s did manufacturers begin to use the piezoelectric effect in industrial detection applications. Since then, this measurement principle has been increasingly used and has become a mature technology with excellent inherent reliability. They have been used successfully in various applications, such as medical, aerospace, nuclear instruments and as a tilt sensor in consumer electronics or a pressure sensor in touch pads on mobile phones. In the automotive industry, piezoelectric elements are used to control combustion when developing internal combustion engines. The sensors are mounted directly into additional holes in the cylinder head or the spark plug / spark plug is equipped with a built-in miniature piezo sensor.

The rise of piezoelectric technology is directly related to a set of inherent advantages. The high modulus of elasticity of many piezoelectric materials is comparable to that of many metals and reaches 10^6 N / m^2. Although piezoelectric sensors are electromechanical systems that react to compression, the sensor elements show almost zero deviation. This provides robustness to piezoelectric sensors, extremely high natural frequency, and excellent linearity over a wide range of amplitude. In addition, piezoelectric technology is insensitive to electromagnetic fields and radiation, allowing measurements to be made in adverse conditions. Some materials used (especially gallium phosphate or tourmaline) are extremely stable at high temperatures, allowing the sensors to have a working range of up to 1000 ° C. Tourmaline displays pyroelectric electricity in addition to the piezoelectric effect; This is the ability to generate an electrical signal when the temperature of the crystal changes. TThis effect is also common to piezo ceramic materials.

A disadvantage of piezoelectric sensors is that they cannot be used for truly static measurements. A static force results in a fixed amount of charge on the piezoelectric material. In conventional reading electronics, imperfect insulating materials and reduced sensor internal resistance cause constant loss of electrons and produce a decreasing signal. High temperatures cause a further drop in internal resistance and sensitivity. The main effect on the piezoelectric effect is that with increasing pressure loads and temperature, sensitivity is reduced due to the formation of twins. While quartz sensors must be cooled during measurements to temperatures above 300 ° C, special types of crystals such as gallium phosphate GaPO4 do not show twin formation up to the melting point of the material itself. However, it is not true that piezoelectric sensors can only be used for very fast processes or in ambient conditions. In fact, many piezoelectric applications produce quasi-static measurements, and other applications operate at temperatures above 500 ° C. Piezoelectric sensors can also be used to determine aromas in air by simultaneously measuring resonance and capacitance. Computer controlled electronics greatly increases the range of potential applications for piezoelectric sensors. Piezoelectric sensors are also seen in nature. Collagen in bone is piezoelectric, and some think that it acts as a biological force sensor.

4.4.3 Principle of Operation

The way a piezoelectric material is cut produces three main operational modes:

Transverse effect: A force applied along a neutral axis (y) generates loads along the (x) direction, perpendicular to the line of force. The amount of charge (C_x) depends on the geometric dimensions of the respective piezoelectric element. When the dimensions a, b, c,

$$C_x = d_{xy} F_y \, b/a \qquad (4.4)$$

Where, a is the dimension in line with the neutral axis, b is in line with the load generating axis and d is the corresponding piezoelectric coefficient.

Longitudinal effect: The amount of charge produced is strictly proportional to the applied force and independent of the size and shape of the piezoelectric element. Placing multiple elements mechanically in series and electrically in parallel is the only way to increase the load output. The resulting load is

$$C_x = d_{xx} F_x \, n \qquad (4.5)$$

Where d_{xx} is the piezoelectric coefficient for a load in the x direction released by forces applied along the x direction (in pC / N). Fx is the Force applied in the x [N] direction and n corresponds to the number of elements stacked.

Shear effect: The charges produced are strictly proportional to the applied forces and independent of the element size and shape. For elements mechanically in series and electrically in parallel the charge is

$$C_x = 2d_{xx}F_x \, n \qquad (4.6)$$

In contrast to the longitudinal and shear effects, the transverse effect allows adjusting the sensitivity in the applied force and the dimension of the element.

4.4.4 Electrical Properties

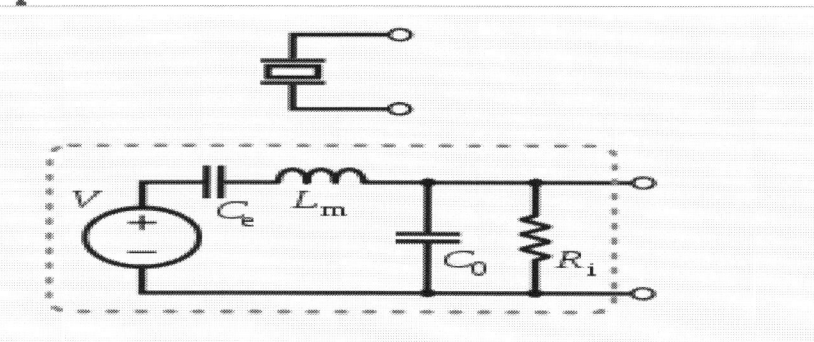

Figure 4.9: Schematic symbol and electronic model of a piezoelectric sensor

A piezoelectric transducer has a very high DC output impedance and can be modeled as a proportional voltage source and a filter network. The voltage V at the source is directly proportional to the applied force, pressure, or voltage. The output signal is related to this mechanical force as if it had passed through the equivalent circuit.

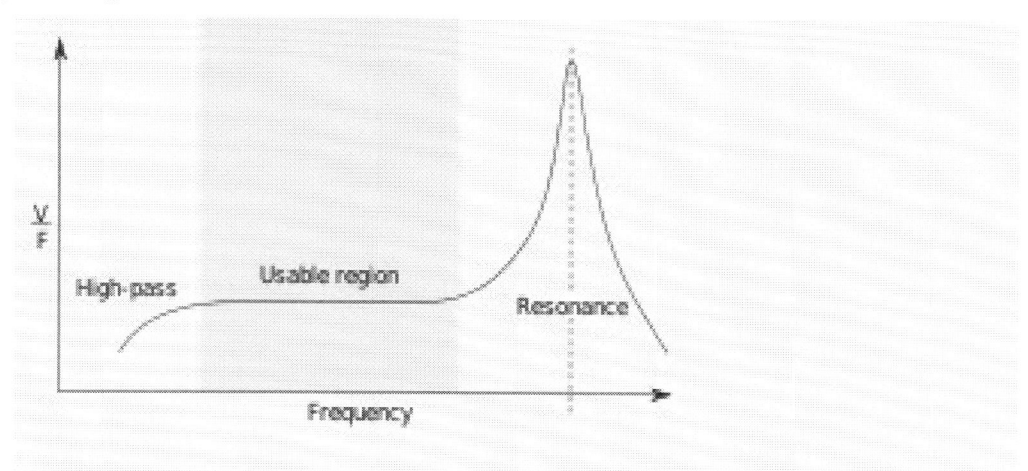

Figure 4.10: Frequency response of a piezoelectric sensor

A detailed model includes the effects of mechanical sensor construction and other non-idealities. The inductance Lm is due to the seismic mass and the inertia of the sensor itself. Ce is inversely proportional to the mechanical elasticity of the sensor. C0 represents the static capacitance of the transducer, resulting from an inertial mass of infinite size. Ri is the insulation leakage resistance of the transducer element. If the sensor is connected to a load resistor, this also works in parallel with the insulation resistance, increasing the high pass cutoff frequency.

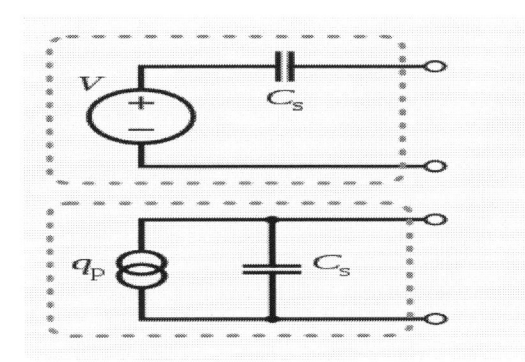

Figure 4.11: sensor modeled as a voltage source in series with the sensor's capacitance or a charge source in parallel with the capacitance

In the flat region, the sensor can be modeled as a voltage source in series with the capacitance of the sensor or a load source in parallel with the capacitance. For use as a sensor, the flat region of the frequency response graph is typically used, between the high-pass cutoff and the resonant peak. The resistance to load and leakage must be large enough so that the low frequencies of interest are not lost. A simplified equivalent circuit model can be used in this region, where Cs represents the capacitance of the sensor surface, determined by the standard formula for parallel plate capacitance. It can also be modeled as a load source in parallel with the source's capacitance, with the load directly proportional to the applied force, as noted above.

4.4.5 Sensor Design

Based on piezoelectric technology, various physical quantities can be measured; The most common are pressure and acceleration. For pressure sensors, a thin membrane and massive base are used, ensuring that an applied pressure specifically loads the elements in one direction. For accelerometers, a seismic mass is attached to the glass elements. When the accelerometer experiences motion, the invariant seismic mass loads the elements according to Newton's second law of motion. The main difference in the operating principle between these two cases is the way they apply forces to the sensor elements. In a pressure sensor, a thin membrane transfers the force to the elements, while in accelerometers an attached seismic mass applies the forces.

Sensors often tend to be sensitive to more than one physical quantity. Pressure sensors show false signals when exposed to vibrations. Therefore, sophisticated pressure sensors use acceleration compensation elements in addition to pressure sensor elements. By carefully pairing those elements, the acceleration signal (released from the compensation element) is subtracted from the combined pressure and acceleration signal to derive the true pressure information. Vibration sensors can also harvest wasted energy from mechanical vibrations. This is accomplished through the use of piezoelectric materials to convert mechanical stress into usable electrical energy.

4.4.6 Materials used for the Piezoelectric Transducers

Two main groups of materials are used for piezoelectric sensors: piezoelectric ceramic and monocrystalline materials. Ceramic materials (such as PZT ceramics) have a piezoelectric constant / sensitivity that is approximately two orders of magnitude higher than that of natural single crystal materials and can be produced by inexpensive sintering processes. The piezoelectric effect in piezoelectric ceramics is "trained", so its high sensitivity degrades over time. This degradation is highly correlated with the increase in temperature. The least sensitive natural monocrystalline materials (gallium phosphate, quartz, tourmaline) have greater long-term stability, when handled with care, almost unlimited. There are also new commercially available monocrystalline materials, such as lead titanate and magnesium lead niobate (PMN-PT). These materials offer improved sensitivity over PZT but have a lower maximum operating temperature and are currently more expensive to manufacture.

There are various materials that exhibit a piezoelectric effect as mentioned above. Materials used for the measurement purpose should possess desirable properties such as stability, high performance, insensitive to extreme temperature and humidity, and ability to be formed or machined in any way. But none of the materials that exhibit piezoelectric effect have all the properties. Quartz, which is a natural crystal, is highly stable but the performance obtained from it is very small. It also offers the advantage of measuring parameters that vary very slowly since they have very low leakage when used with high input impedance amplifiers. Due to its stability, quartz is commonly used in piezoelectric transducers. It is usually cut into a rectangular or square plate shape and held between two electrodes. The glass is connected to the appropriate electronic circuit to obtain a sufficient output. Rochelle salt, a synthetic crystal, provides the highest performance among all materials that exhibit a piezoelectric effect. However, it has to be protected from moisture and cannot be used at a temperature above 115 degrees F. In general, synthetic crystals are more sensitive and perform better than natural crystals.

4.4.7 Advantages of Piezoelectric Transducers

Every device has certain advantages and limitations. The piezoelectric transducers offer several advantages as mentioned below:

1) **High frequency response:** They offer very high frequency response that means the parameter changing at very high speeds can be sensed easily.

2) **High transient response:** The piezoelectric transducers can detect the events of microseconds and also give the linear output.
3) **High output:** They offer high output that be measured in the electronic circuit.
4) The piezoelectric transducers are small in size and have rugged construction.

4.4.8 Limitations of Piezoelectric Transducers

Some of the limitations of piezoelectric transducers are:

1) **The output is low:** the output obtained from the piezoelectric transducers is low, so an external electronic circuit must be connected.

2) **High impedance:** Piezoelectric crystals have high impedance, so they must be connected to the amplifier and auxiliary circuit, which have the potential to cause measurement errors. To reduce these errors, high input impedance amplifiers and long cables should be used.

3) **Forming:** it is very difficult to give the desired shape to the crystals with sufficient resistance.

4.4.9 Applications of the Piezoelectric Transducers

1) Piezoelectric transducers are most useful for dynamic measurements, that is, parameters that change at a fast speed. This is because the instrument does not maintain the developed potential under static conditions. Therefore, piezoelectric crystals are mainly used to measure quantities such as surface roughness, and also in accelerometers and vibration pads.

2) For the same reasons, they can be used to study high-speed phenomena such as explosions and blast waves. They are also used in aerodynamic shock tube and seismograph work (used to measure acceleration and vibration in rockets).

3) Often times piezoelectric sensors or transducers are used in conjunction with strain gauges to measure force, stress, vibrations, etc.

4) Automotive companies used piezoelectric transducers to detect knocks in engine blocks.

5) Piezoelectric transducers are used in medical treatment, son's chemistry, and industrial processing equipment to control power.

4.5 Hall Effect Sensor

A Hall effect sensor is a transducer that varies its output voltage in response to a magnetic field. Hall effect sensors are used for proximity detection, positioning, speed detection and current detection applications. In its simplest form, the sensor works like an analog transducer, directly returning a voltage. With a known magnetic field, its distance from the Hall plate can be determined. Using groups of sensors, the relative position of the magnet can be deduced. Often, a Hall sensor is combined with threshold detection to act as a switch. Commonly seen in industrial applications such as the illustrated pneumatic cylinder, they are also used in consumer equipment; for example, some computer printers use them to detect lost paper and open covers. When high reliability is required, they are used on keyboards. Hall sensors are commonly used to measure wheel and axle speed, such as internal combustion engine ignition timing, tachometers, and anti-lock brake systems. They are used in brushless DC electric motors to detect the position of the permanent magnet. On the illustrated wheel with two equally spaced magnets, the sensor voltage will peak twice for each revolution. This arrangement is commonly used to regulate the speed of disk drives.

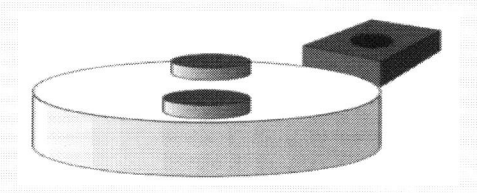

Figure 4.12: A wheel containing two magnets passing by a Hall Effect sensor

4.5.1 Hall Probe

A Hall probe contains a semiconductor crystal composed of indium, such as indium antimonite, mounted on an aluminum backing plate and encapsulated in the probe head. The plane of the glass is perpendicular to the probe handle. The glass connection cables are lowered through the handle to the circuit box. When the Hall probe is held so that the magnetic field lines pass at a right angle through the probe sensor, the meter gives a reading of the value of the magnetic flux density (B). A current is passed through the crystal which, when placed in a magnetic field, has a "Hall effect" voltage developed through it. The Hall effect is seen when a conductor passes through a uniform magnetic field. The drift of natural electrons from charge carriers causes the magnetic field to apply a Lorentz force (the force exerted on a charged particle in an electromagnetic field) to these charge carriers. The result is what looks like a charge gap, with a buildup of positive or negative charges at the bottom or top of the board. The glass measures 5 mm square. The probe handle, which is made of non-ferrous material, has no disturbing effect in the field. A Hall probe must be calibrated with a known value of magnetic field strength. For a solenoid, the Hall probe is placed in the center.

4.5.2 Hall Effect Sensor Principals

When a beam of charged particles passes through a magnetic field, the forces act on the particles and the beam deviates from a straight path. The flow of electrons through a conductor is known as a charged carrier beam. When a conductor is placed in a magnetic field perpendicular to the direction of the electrons, they will deviate from a straight path. As a consequence, one plane of the conductor will be negatively charged and the opposite side will be positively charged. The voltage between these planes is called the Hall voltage. When the force on the charged particles of the electric field balances the force produced by the magnetic field, the separation of them will stop. If the current is not changing, then the Hall voltage is a measure of the magnetic flux density. Basically there are two types of Hall effect sensors. One is linear, which means that the voltage output is linearly dependent on the magnetic flux density; the other is called a threshold, which means that there will be a sharp decrease in the output voltage at each magnetic flux density.

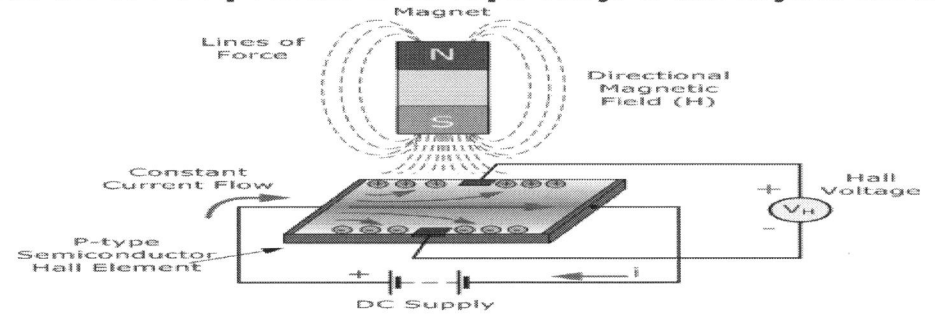

Figure 4.13: Hall Effect Sensor Principals

Hall effect sensors basically consist of a thin piece of p-type rectangular semiconductor material such as gallium arsenide (GaAs), indium antimonide (InSb), or indium arsenide (InAs) that passes a direct current through itself. When the device is placed within a magnetic field, the magnetic flux lines exert a force on the semiconductor material that deflects the charge carriers, electrons, and holes on either side of the semiconductor slab. This movement of the charge carriers is the result of the magnetic force they experience as they pass through the semiconductor material. As these electrons and holes move sideways, a potential difference is produced between the two sides of the semiconductor material by the accumulation of these charge carriers. Then, the movement of electrons through the semiconductor material is affected by the presence of an external magnetic field that is at right angles to it and this effect is greater in a flat material of rectangular shape.

The effect of generating a measurable voltage by using a magnetic field is called the Hall Effect after Edwin Hall discovered it in the 1870s with the basic physical principle underlying the Hall effect which is the Lorentz force. To generate a potential difference across the device, the magnetic flux lines must be perpendicular, (90o) to the current flux and have the correct polarity, usually a south pole. The Hall effect provides information on the type of magnetic pole and the magnitude of the magnetic field. For example, a south pole would cause the device to produce a voltage output, while a north pole would have no effect. In general, Hall effect sensors and switches are designed to be "OFF" (open circuit condition) when no magnetic field is present. They only "turn ON" (closed circuit condition) when subjected to a magnetic field of sufficient strength and polarity.

4.5.3 Hall Effect Magnetic Sensor

The output voltage, called Hall voltage (VH) of the basic Hall element is directly proportional to the intensity of the magnetic field passing through the semiconductor material (output $\propto$ H). This output voltage can be quite small, just a few microvolts, even when subjected to strong magnetic fields, so most commercially available Hall effect devices are built with built-in DC amplifiers, logic switching circuits, and regulators. Voltage to improve sensitivity, hysteresis and sensor output voltage. This also enables the Hall Effect Sensor to operate on a wider range of power supplies and magnetic field conditions.

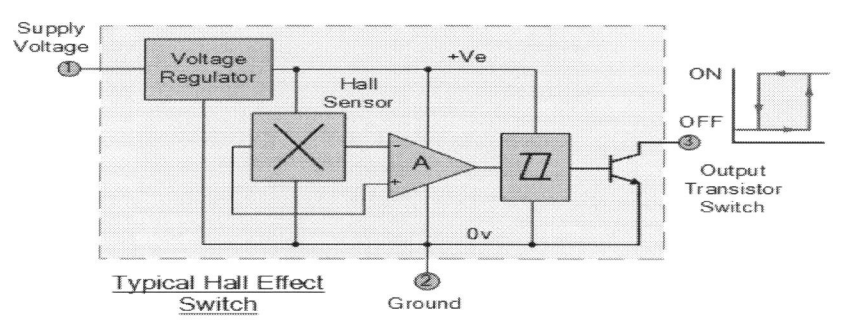

Figure 4.14: The Hall Effect Magnetic Sensor

Hall effect sensors are available with either linear or digital outputs. The output signal for linear (analog) sensors is taken directly from the output of the op-amp, the output voltage being directly proportional to the magnetic field passing through the Hall sensor. This Hall voltage output is given as:

$$V_H = R_H \left(\frac{I}{t} \times B \right) \tag{4.7}$$

Where,

V_H is the Hall Voltage in volts

R_H is the Hall Effect co-efficient

I is the current flow through the sensor in amps

t is the thickness of the sensor in mm

B is the Magnetic Flux density in Tesla

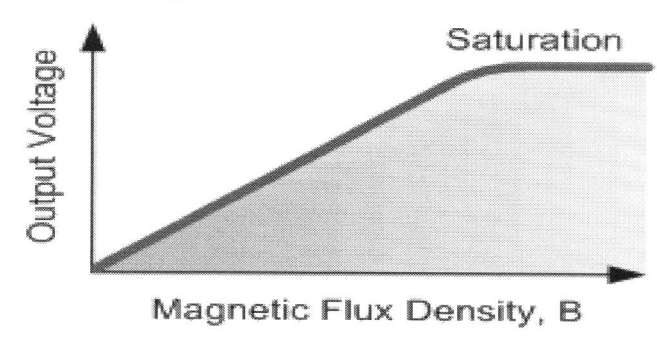

Figure 4.15: Output Voltage versus Magnetic Flux Density

Linear or analog sensors provide a continuous voltage output that increases with a strong magnetic field and decreases with a weak magnetic field. In linear output Hall effect sensors, as the magnetic field strength increases, the amplifier's output signal will also increase until it begins to become saturated by the limits imposed by the power supply. Any further increase in the magnetic field will have no effect on the output, but will lead more to saturation. Digital output sensors, on the other hand, have a Schmitt trigger with built-in hysteresis connected to the op amp. When the magnetic flux passing through the Hall sensor exceeds a preset value, the output of the device quickly switches between its "OFF" conditions to an "ON" condition without any contact bounce. This built-in hysteresis eliminates any oscillation of the output signal as the sensor enters and leaves the magnetic field. So, the digital output sensors have only two states, "ON" and "OFF".

There are two basic types of Hall effect digital sensor, bipolar and unipolar. Bipolar sensors require a positive magnetic field (South Pole) to operate and a negative field (North Pole) to release them, while unipolar sensors require only a single magnetic South Pole to operate and release them as they enter and exit the magnetic. countryside. Most Hall effect devices cannot directly switch large electrical loads since their output drive capacities are very small, around 10 to 20 mA. For large current loads, an open collector NPN transistor (current sink) is added to the output. This transistor works in its saturated region as an NPN sink switch that shortens the output terminal to ground each time the applied flux density is greater than that of the "ON" preset point. The output switching transistor can be an open emitter transistor, an open collector transistor configuration, or both that provide a push-pull output type configuration that can absorb enough current to directly drive many loads, including relays, motors, LEDs and lamps.

4.5.4 Materials for Hall Effect Sensors

The key factor determining the sensitivity of Hall effect sensors is the high electron mobility. As a result, the following materials are especially suitable for Hall effect sensors:

- Gallium Arsenide (GaAs)
- Indium Arsenide (InAs)
- Indium Phosphide (InP)
- Indium Antimonide (InSb)
- Graphene

4.5.5 Signal Processing and Interface

Hall effect sensors are linear transducers. As a result, such sensors require a linear circuit to process the sensor output signal. Such a linear circuit:

- Provides a constant conduction current to the sensors.
- Amplify the output signal

In some cases, the linear circuit can cancel the offset voltage of the Hall effect sensors. Furthermore, AC modulation of the driving current can also reduce the influence of this offset voltage. Hall effect sensors with linear transducers are commonly integrated with digital electronics. This allows advanced corrections of the sensor characteristics (eg, temperature coefficient corrections) and digital interface to microprocessor systems. In some Hall effect sensor IC solutions a DSP is used, which provides more choice between processing techniques. Hall effect sensor interfaces can include input diagnostics, fault protection for transient conditions, and short circuit / open circuit detection. You can also supply and monitor the current to the Hall effect sensor. Precision IC products are available to handle these characteristics.

4.5.6 Advantages

A Hall Effect sensor can function as an electronic switch.

- Such a switch costs less than a mechanical switch and is much more reliable.
- It can be operated up to 100 kHz.
- No contact bounces because a solid state switch with hysteresis is used instead of a mechanical contact.
- It will not be affected by environmental contaminants since the sensor is in a sealed package. Therefore, it can be used in severe conditions.

In the case of the linear sensor (for magnetic field intensity measurements), a Hall effect sensor:

- can measure a wide range of magnetic fields
- is available to measure magnetic fields of the north or south pole
- can be flat

4.5.7 Disadvantages

Hall effect sensors provide much lower measurement accuracy than fluxgate magnetometers or magneto resistance based sensors. In addition, Hall effect sensors move significantly, requiring compensation.

4.5.8 Applications of Hall Effect Sensor

1. Direct current (DC) Transformers

1. Direct current (DC) transformers

Hall effect sensors can be used for non-contact measurements of DC current in current transformers. In such a case, the Hall effect sensor is mounted in the space in the magnetic core around the current conductor. As a result, the DC magnetic flux can be measured and the DC current in the conductor can be calculated.

2. Automotive Fuel Level Gauge

The Hall sensor is used in some automotive fuel level gauges. The main principle of operation of said indicator is the detection of the position of a floating element. This can be done using a vertical floating magnet or a rotary lever sensor.

- In a vertical flotation system, a permanent magnet is mounted on the surface of a floating object. The current carrying conductor is attached to the top of the tank by aligning itself with the magnet. When the fuel level increases, an increasing magnetic field is applied to the current resulting in a higher Hall voltage. As the fuel level decreases, the Hall voltage will also decrease. The fuel level is indicated and displayed by the appropriate Hall voltage signal condition.
- In a rotary lever sensor, a diametrically magnetized ring magnet rotates around a linear hall sensor. The sensor only measures the perpendicular (vertical) component of the field. The strength of the measured field is directly correlated with the angle of the lever and, therefore, with the level of the fuel tank.

3. Keyboard Switch: Developed by Everett A. Vorthmann and Joeseph T. Maupin for Micro Switch (a division of Honeywell) in 1969, the switch was known to still be in production until the late 1990s. The switch is one of the most high. Quality keyboard switches never produced, with reliability being the primary focus of the design. The key switches have been proven to have a lifespan of over 30 billion keystrokes; The switch also has two open collector outputs for increased reliability. The Honeywell Hall Effect switch is the most famous switch used on the Space-cadet keyboard, a keyboard used on LISP machines. There are several possible paths of motion to detect a magnetic field, and here are two of the most common detection settings that use a single magnet: front detection and side detection.

4. Head-on Detection

As its name implies, "frontal detection" requires that the magnetic field be perpendicular to the Hall-effect detection device and that for detection; it approaches the sensor directly towards the active face. A kind of "head-on" approach. This frontal approach generates an output signal, VH, which in linear devices represents the intensity of the magnetic field, the density of the magnetic flux, as a function of the distance from the Hall effect sensor.

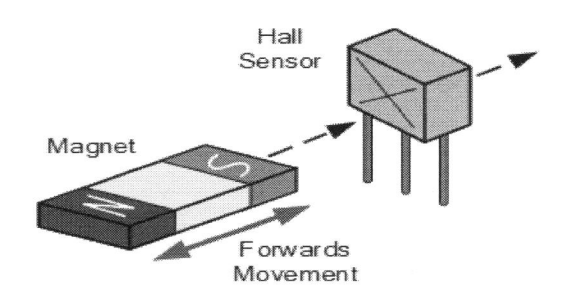

Figure 4.16: Head-on Detection

The closer and therefore the stronger the magnetic field, the higher the output voltage and vice versa. Linear devices can also differentiate between positive and negative magnetic fields. Non-linear devices can be made to activate the "ON" output at a preset air gap distance away from the magnet to indicate position detection.

5. Sideways Detection

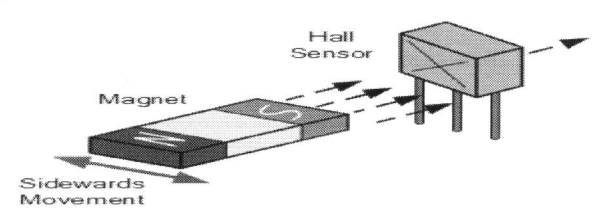

Figure 4.17: Sideways Detection

The second detection setting is "side detection". This requires moving the magnet across the face of the Hall effect element in a lateral motion. Lateral or sliding detection is useful for detecting the presence of a magnetic field as it moves across the face of the Hall element within a fixed air gap distance, for example by counting rotational magnets or the rotational speed of the engines. Depending on the position of the magnetic field as it passes through the centerline of the sensor's zero field, a linear output voltage can be produced that represents both a positive and negative output. This allows detection of directional movement that can be both vertical and horizontal.

There are many different applications for Hall effect sensors, especially as proximity sensors. They can be used in place of light and optical sensors where environmental conditions consist of water, vibration, dirt, or oil, such as in automotive applications. Hall effect devices can also be used for current detection. We know from previous tutorials that when a current passes through a conductor, a circular electromagnetic field is produced around it. By placing the Hall sensor next to the driver, electrical currents from a few milliamps to thousands of amps can be measured from the generated magnetic field without the need for large or expensive transformers and coils.

In addition to detecting the presence or absence of magnets and magnetic fields, Hall effect sensors can also be used to detect ferromagnetic materials such as iron and steel by placing a small permanent "polarization" magnet behind the active area of the device. The sensor is now in a permanent and static magnetic field, and any change or disturbance of this magnetic field by the introduction of a ferrous material will be detected with sensitivities as low as possible mV / G. There are many different ways to connect the sensors. Hall effect to electrical and electronic circuits, depending on the type of device, either digital or linear. A very simple and easy example to build is to use a light emitting diode as shown below.

6. Positional Detector

Detecting the presence of magnetic objects (connected with position detection) is the most common industrial application of Hall effect sensors, especially those that operate in the switch mode (on / off mode). Hall effect sensors are also used in the brushless DC motor to detect the rotor position and change transistors in the correct sequence. Smartphones use hallway sensors to determine if the Flip Cover accessory is closed.

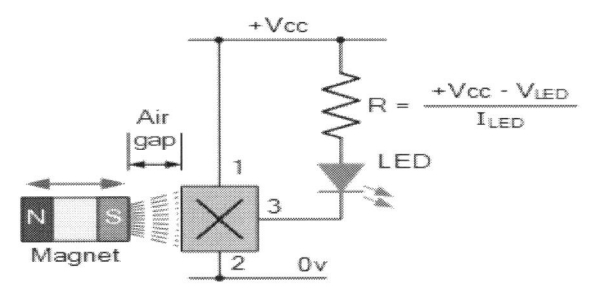

Figure 4.18: Positional Detector

This front positional detector will be "OFF" when no magnetic field is present (0 gauss). When the permanent magnets of the South Pole (positive gauss) move perpendicular to the active area of the Hall effect sensor, the device turns on and the LED turns on. Once turned on, the Hall effect sensor remains on. To turn off the device and thus the "OFF" LED, the magnetic field must be reduced below the release point of the unipolar sensors or exposed to a magnetic north pole (negative gauss) for the bipolar sensors. The LED can be replaced with a larger power transistor if the Hall effect sensor output is required to change larger current loads.

4.6 Photoelectric Sensor

A photoelectric sensor, or photoelectric eye, is equipment used to discover the distance, absence, or presence of an object through the use of a light transmitter, often infrared, and a photoelectric receiver. They are widely used in industrial manufacturing. There are three different useful types: opposed (through beam), retro-reflective, and proximity-sensing (diffused).

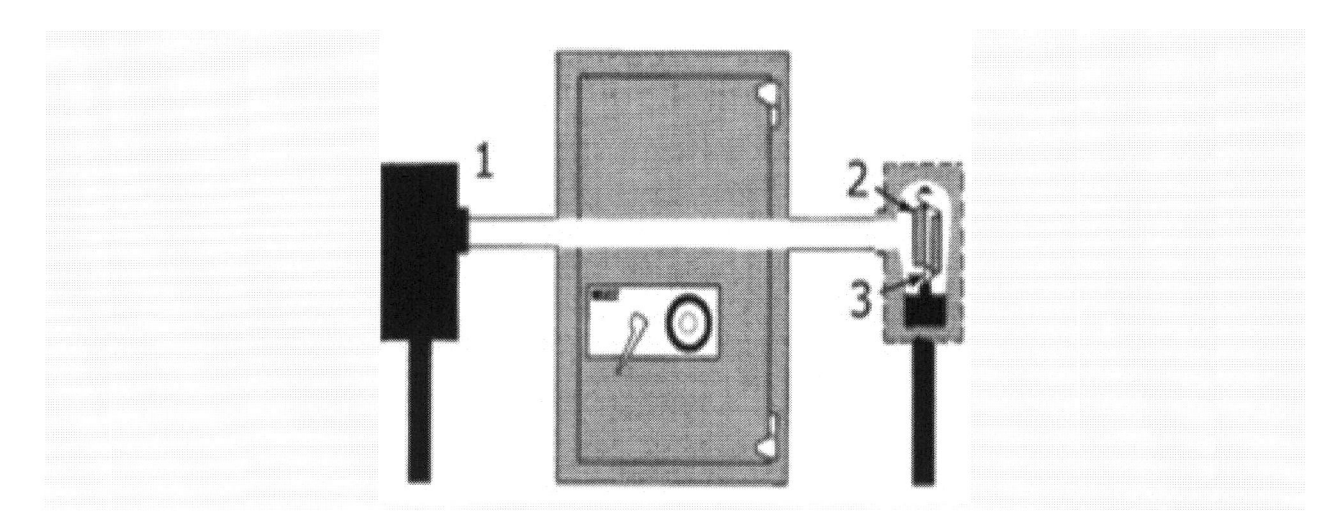

Figure 4.19: Conceptual through-beam system to detect unauthorized access to a secure door

4.6.1 Types of Photoelectric Sensor

A self-contained photoelectric sensor contains the optics, along with the electronics. It only requires a power source. The sensor performs its own modulation, demodulation, amplification, and output switching. Some autonomous sensors provide options such as built-in control timers or counters. Due to technological progress, autonomous photoelectric sensors have become smaller and smaller. Remote photoelectric sensors used for remote sensing contain only the optical components of a sensor. The circuits for power input, amplification, and output switching are located elsewhere, usually on a control panel. This allows the sensor, itself, to be very small. Also, the controls for the sensor are more accessible as they can be larger. When space is restricted or the environment is too hostile even for remote sensors, fiber optics can be used. Optical fibers are passive components of mechanical detection. They can be used with remote or autonomous sensors. They have no electrical circuits or moving parts, and can safely channel light into and out of harsh environments.

4.6.2 Sensing Modes

A continuous beam arrangement consists of a receiver located within the line of sight of the transmitter. In this mode, an object is detected when the light beam cannot reach the receiver from the transmitter. A retro reflective arrangement places the transmitter and receiver in the same location and uses a reflector to bounce the light beam from the transmitter to the receiver. An object is detected when the beam is interrupted and does not reach the receiver. A proximity (diffuse) detection arrangement is one in which the transmitted radiation must be reflected back to the object to reach the receiver. In this mode, an object is detected when the receiver sees the transmitted source rather than when it cannot see it. Just like retroreflective sensors, diffuse sensor emitters and receivers are located in the same housing. But the lens acts as a reflector, so that light detection reflects off the object of the disturbance. The emitter emits a beam of light (usually pulsed infrared, visible red, or laser) that diffuses in all directions, filling a detection area. The target then enters the area and deflects part of the beam toward the receiver. Detection occurs and the output turns on or off when enough light falls on the receiver.

Some photographic eyes have two different types of operation, light and dark. Light photoelectric eyes work when the receiver "receives" the signal from the transmitter. The dark eyes that work with photographs become operational when the receiver "does not receive" the signal from the transmitter. The detection range of a photoelectric sensor is its "field of view", or the maximum distance from which the sensor can retrieve information, minus the minimum distance. A minimum detectable object is the smallest object that the sensor can detect. The most accurate sensors can often have tiny, detectable minimum objects.

Figure 4.20: Smoke detector use a photoelectric sensor to warn of smoldering fires.

4.6.3 Photoelectric Effect

The photoelectric effect or photoemission is the production of electrons or other free carriers when light shines on a material. The electrons emitted in this way can be called photoelectrons. The phenomenon is commonly studied in electronic physics, as well as in fields of chemistry, such as quantum chemistry or electrochemistry. According to classical electromagnetic theory, this effect can be attributed to the transfer of energy from light to an electron. From this perspective, an alteration in light intensity would induce changes in the electron emission rate of the metal. Furthermore, according to this theory, sufficiently dim light would be expected to show a time lag between the initial brightness of its light and the subsequent emission of an electron. However, the experimental results did not correlate with either of the two predictions made by classical theory.

Difference between Modes

Instead, electrons are dislodged only by the impact of photons when those photons reach or exceed a threshold frequency (energy). Below that threshold, the metal does not emit electrons, regardless of light intensity or time of light exposure. To make sense of the fact that light can eject electrons even if its intensity is low, Albert Einstein proposed that a beam of light is not a wave that propagates through space, but rather a collection of discrete wave packets (photons), each with hf energy. This sheds light on Max Planck's earlier discovery of the Planck relation (E = hf) that unites energy (E) and frequency (f) as a result of quantifying energy.

Name	Advantages	Disadvantages
Through-Beam	- Most accurate - Longest sensing range - Very reliable	- Must install at two points on system: emitter and receiver - Costly - must purchase both emitter and receiver
Reflective	- Only slightly less accurate than through-beam - Sensing range better than diffuse - Very reliable	- Must install at two points on system: sensor and reflector - Slightly more costly than diffuse - Sensing range less than through-beam
Diffuse	- Only install at one point - Cost less than through-beam or reflective	- Less accurate than through-beam or reflective - More setup time involved

The factor h is known as the Planck constant. In 1887, Heinrich Hertz discovered that electrodes illuminated with ultraviolet light create electrical sparks more easily. In 1905, Albert Einstein published an article explaining experimental data on the photoelectric effect as a result of the energy of light carried in quantized discrete packages. This discovery led to the quantum revolution. In 1914, Robert Millikan's experiment confirmed Einstein's law on the photoelectric effect. Einstein was awarded the Nobel Prize in 1921 for "his discovery of the law of the photoelectric effect" and Millikan received the Nobel Prize in 1923 for "his work on the elementary charge of electricity and the photoelectric effect". The photoelectric effect requires photons with energies close to zero (in the case of negative electronic affinity) at more than 1 MeV for central electrons in elements with a high atomic number. Emission of conduction electrons from typical metals generally requires a few electron volts, which correspond to short wavelength ultraviolet or visible light. The study of the photoelectric effect led to important steps to understand the quantum nature of light and electrons and influenced the formation of the concept of wave-particle duality. Other phenomena in which light affects the movement of electrical charges include the photoconductive effect (also known as photoconductivity or photo resistivity), the photovoltaic effect, and the photo electrochemical effect.

Photoemission can occur from any material, but it is more easily observable from metals or other conductors because the process produces a load imbalance, and if this load imbalance is not neutralized by the current flow (enabled by conductivity), the potential emission barrier increases until the emission current ceases. It is also common to have the emitting surface in a vacuum, since the gases impede the flow of the photoelectrons and make their observation difficult. Furthermore, the energy barrier for photoemission generally increases with thin oxide layers on metal surfaces if the metal has been exposed to oxygen, so most practical experiments and devices based on the photoelectric effect use clean metal surfaces in the empty. When photoelectron is emitted in a solid rather than in a vacuum, the term internal photoemission is often used, and vacuum emission is distinguished as external photoemission.

4.6.4 Emission Mechanism

The photons in a light beam have a characteristic energy proportional to the frequency of light. In the photoemission process, if an electron inside a material absorbs the energy of a photon and acquires more energy than the work function (the energy of electron bonding) of the material, it is ejected. If the energy of the photon is too low, the electron cannot escape from the material. Since an increase in the intensity of low-frequency light will only increase the number of low-energy photons sent during a given time interval, this change in intensity will not create any photons with enough energy to dislodge an electron. Therefore, the energy of the emitted electrons does not depend on the intensity of the incoming light, but only on the energy (equivalent frequency) of the individual photons. It is an interaction between the incident photon and the outermost electrons. Electrons can absorb energy from photons when they are irradiated, but generally follow an "all or nothing" principle.

4.6.5 Experimental Observations of Photoelectric Emission

The theory of the photoelectric effect must explain experimental observations of the emission of electrons from an illuminated metal surface. For a given metal, there is a certain minimum frequency of incident radiation below which no photoelectrons are emitted. This frequency is called the threshold frequency. By increasing the frequency of the incident beam, keeping the number of incident photons fixed (this would result in a proportional increase in energy) increases the maximum kinetic energy of the emitted photoelectrons. Therefore, the stop voltage increases. The number of electrons also changes because the probability that each photon produces an emitted electron is a function of the energy of the photon. If the intensity of the incident radiation of a given frequency is increased, there is no effect on the kinetic energy of each photoelectron.

Above the frequency threshold, the maximum kinetic energy of the emitted photoelectron depends on the frequency of the incident light, but is independent of the intensity of the incident light as long as the latter is not too high. For a given metal and the frequency of the incident radiation, the rate at which the photoelectrons are ejected is directly proportional to the intensity of the incident light.

An increase in the intensity of the incident beam (keeping the frequency fixed) increases the magnitude of the photoelectric current, although the stopping voltage remains the same.

The time lapse between the incidence of radiation and the emission of a photoelectron is very small, less than 10−9 seconds. The direction of distribution of the emitted electrons peaks in the polarization direction (the direction of the electric field) of the incident light, if it is linearly polarized.

Stopping potential

The relationship between current and applied voltage illustrates the nature of the photoelectric effect. For discussion, a light source illuminates a P plate, and another Q plate electrode collects the emitted electrons. We vary the potential between P and Q and measure the current flowing in the external circuit between the two plates.

Work function and Cut off Frequency

If the frequency and intensity of the incident radiation are fixed, the photoelectric current gradually increases with an increase in the positive potential at the collecting electrode until all the emitted photoelectrons are collected. The photoelectric current reaches a saturation value and does not increase further for any increase in positive potential. The saturation current increases with increasing light intensity. It also increases with higher frequencies due to a higher probability of electron emission when collisions with higher energy photons occur.

$$E = \frac{hc}{\lambda} \tag{4.8}$$

Where, E is the work function and λ is the cut-off wavelength. If we apply a negative potential to the collector plate Q with respect to the P plate and gradually increase it, the photoelectric current decreases, becoming zero at a certain negative potential. The negative potential in the collector at which the photoelectric current becomes zero is called the stop potential or cut potential. We see that the stop voltage varies linearly with the frequency of light, but it depends on the type of material. For any particular material, there is a threshold frequency that must be exceeded, regardless of light intensity, to observe any electron emission.

Three-step model

In the X-ray regime, the photoelectric effect on the crystalline material is often broken down into three steps: internal photoelectric effect. The hole left can lead to the Auger effect, which is visible even when the electron does not leave the material. In molecular solids, the phonons are excited in this step and can be seen as lines in the final electronic energy. The internal photographic effect must have a dipole allowed. The transition rules for atoms are translated through the tight bond pattern in the crystal. They are similar in geometry to plasma oscillations in that they have to be transverse.

- Ballistic transport of half the electrons to the surface. Some electrons are scattered.
- Electrons escape from the material on the surface.

In the three-step model, an electron can take multiple paths through these three steps. All the ways can interfere in the sense of the integral formulation of the way. For surface states and molecules, the three-step model still makes sense, since even most atoms have multiple electrons that can scatter the leaving electron.

4.7 Light Sensors

A light sensor generates an output signal that indicates the intensity of the light by measuring the radiant energy that exists in a very narrow range of frequencies basically called "light", and that varies in frequency from "infrared" to "visible" up to " Ultraviolet "light spectrum.

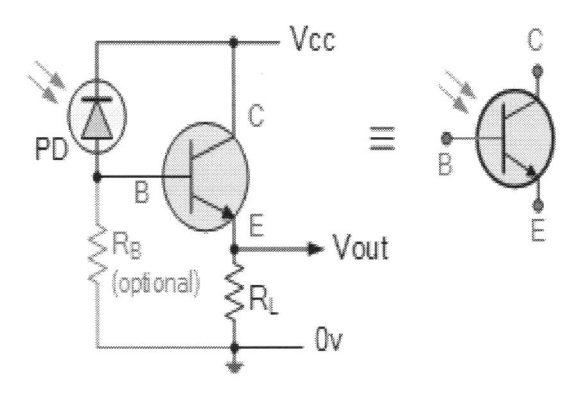

Figure 4.20: Light Sensors

The light sensor is a passive device that converts this "light energy" either visible or in the infrared parts of the spectrum into an electrical signal output. Light sensors are more commonly known as "photoelectric devices" or "photosensors" because they convert the energy of light (photons) into electricity (electrons). Photoelectric devices can be grouped into two main categories, those that generate electricity when illuminated, such as photovoltaics or photoemitters, etc., and those that change their electrical properties in any way, such as photographic resistors or photoconductors. This leads to the following device classification.

- **Photoemitting cells:** they are photographic devices that release free electrons from a light-sensitive material such as cesium when they are hit by a photon of sufficient energy. The amount of energy photons have depends on the frequency of light, and the higher the frequency, the more energy photons have to convert the energy of light into electrical energy.
- **Photoconductive cells:** These photographic devices vary their electrical resistance when subjected to light. Photoconductivity results from light hitting a semiconductor material that controls the flow of current through it. Therefore, more light increases the current for a given applied voltage. The most common photoconductive material is cadmium sulfide used in LDR photocells.

- **Photovoltaic cells:** These photographic devices generate an emf in proportion to the received radiant light energy and have an effect similar to photoconductivity. The light energy falls on two semiconductor materials sandwiched together creating a voltage of approximately 0.5V. The most common photovoltaic material is selenium used in solar cells.
- **Photo junction devices:** These photographic devices are mainly true semiconductor devices, like photodiode or phototransistor, which use light to control the flow of electrons and holes through their PN junction. Photo bonding devices are specifically designed for detector application and light penetration with their spectral response adjusted to the wavelength of incident light.

4.7.1 The Photoconductive Cell

A photoconductive light sensor does not produce electricity, but simply changes its physical properties when subjected to light energy. The most common type of photoconductive device is photographic resistance that changes its electrical resistance in response to changes in light intensity. Photoresistors are semiconductor devices that use light energy to control the flow of electrons and, therefore, the current flowing through them. The commonly used photoconductor cell is called the light dependent resistor or LDR.

Light Dependent Resistor (LDR)

Light Dependent Resistance (LDR) is made from a piece of exposed semiconductor material like cadmium sulfide that changes its electrical resistance from several thousand ohms in the dark to just a few hundred ohms when light falls on it creating pairs of electrons in the hole. material. The net effect is an improvement in your conductivity with a decrease in resistance for an increase in illumination. Furthermore, photoresist cells have a long response time that requires many seconds to respond to a change in light intensity.

Figure 4.22: Typical Light Dependent Resistor (LDR)

Materials used as a semiconductor substrate include lead sulfide (PbS), lead selenide (PbSe), indium antimonide (InSb) that detect light in the infrared range, the most commonly used photoresist light sensor being cadmium sulfide (Cds)) Cadmium sulfide is used in the manufacture of photoconductive cells because its spectral response curve closely matches that of the human eye and can even be controlled using a simple torch as a light source. Typically then, it has a maximum sensitivity wavelength (λp) of about 560nm to 600nm in the visible spectral range.

The Light Dependent Resistor Cell

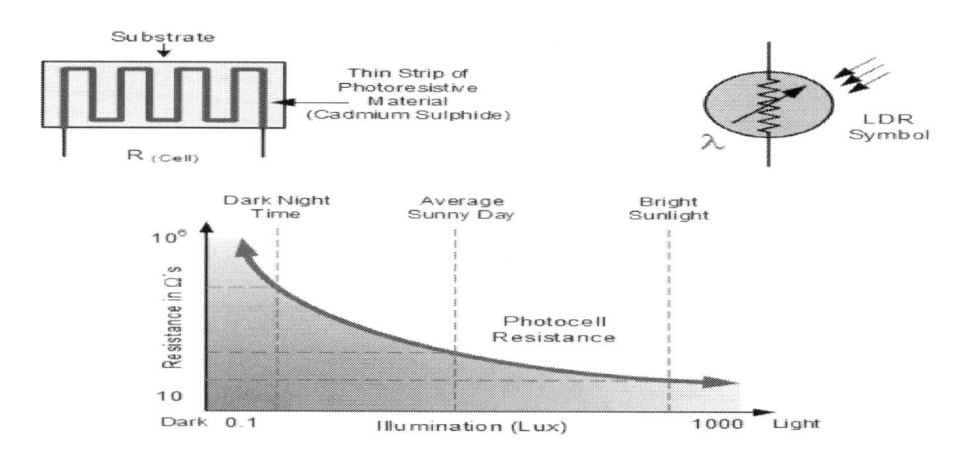

The most widely used photo resist light sensor is the cadmium sulfide photoconductor cell ORP12. This light dependent resistance has a spectral response of approximately 610 nm in the region of light yellow to orange. The resistance of the cell when it is not illuminated (dark resistance) is very high, of approximately 10MΩ, which drops to approximately 100Ω when it is fully illuminated (resistance on). To increase the dark resistance, and therefore reduce the dark current, the resistive path forms a zigzag pattern through the ceramic substrate. The CdS photocell is a very low-cost device that is often used in automatic dimming, darkness or twilight detection to turn street lights on and off and for photographic exposure meter type applications. Connecting a light dependent resistor in series with a standard resistor like this through a single DC supply voltage has a major advantage; A different voltage will appear at your junction for different light levels. The amount of voltage drop across the series resistor, R_2, is determined by the resistive value of the light-dependent resistor, RLDR. This ability to generate different voltages produces a very useful circuit called a "potential divider" or voltage divider network.

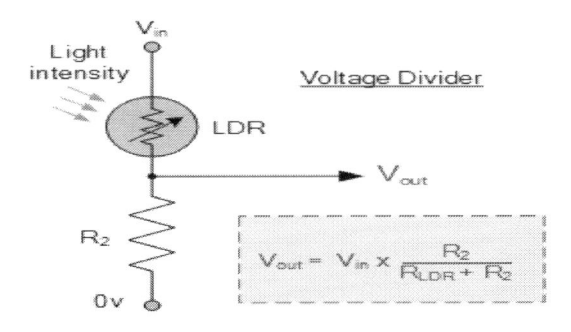

$$V_{out} = V_{in} \times \frac{R_2}{R_{LDR} + R_2}$$

Figure 4.24: Photographic exposure meter type Applications

As we know, the current through a series circuit is common and as the LDR changes its resistive value due to the intensity of the light, the voltage present in V_{OUT} will be determined by the voltage divider formula. The resistance of an LDR, RLDR can range from approximately 100Ω in sunlight, to over 10MΩ in absolute darkness with this resistance variation converted to a voltage variation in V_{OUT} as shown. A simple use of a light dependent resistor is as a light sensitive switch as shown below.

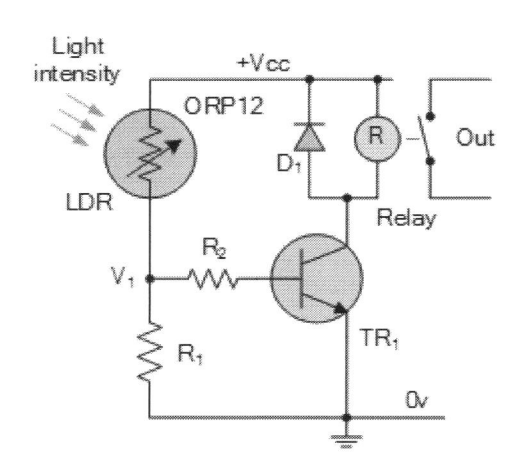

Figure 4.25: LDR Switch

This basic light sensor circuit is of a relay output light activated switch. A potential divider circuit is formed between the photoresistor, LDR, and resistor R1. When no light is present, i.e. in the dark, the resistance of the LDR is very high in the Mega ohm (MΩ) range, so zero-base bias is applied to transistor TR1 and the relay turns off or is "TURNS OFF". As the light level increases, the resistance of the LDR begins to decrease, causing the base bias voltage at V1 to increase. At some point determined by the potential divider network formed with resistor R1, the base bias voltage is high enough to turn on transistor TR1 and activate the relay, which in turn is used to control some external circuits. As the light level falls back into darkness again, the resistance of the LDR increases, causing the base voltage of the transistor to decrease, turning off the transistor and relay at a fixed light level, again determined by the network potential divide. resistor R1 with a potentiometer VR1, the point at which the relay turns on or off can be preset to a particular light level. This type of simple circuit shown above has a fairly low sensitivity and its switching point may not be consistent due to variations in temperature or supply voltage. A more sensitive precision light activated circuit can be easily made by incorporating the LDR in a "Wheatstone Bridge" arrangement and replacing the transistor with an op amp as shown.

Light Level Sensing Circuit

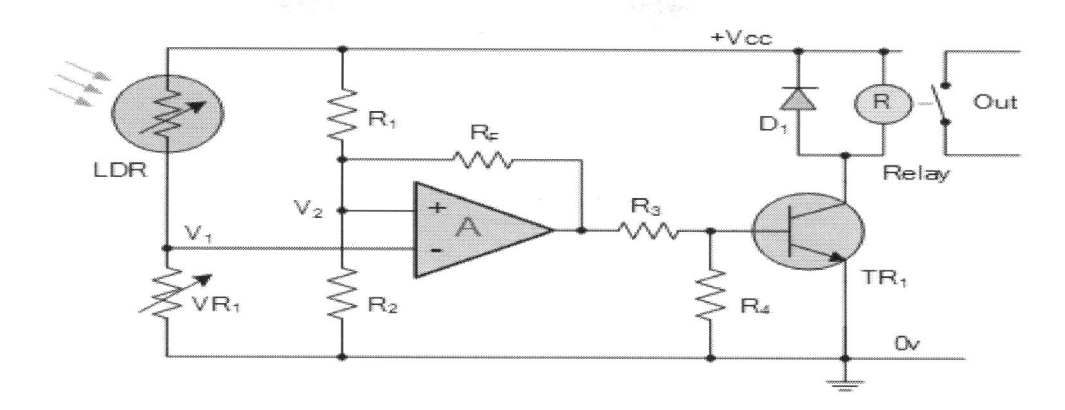

Figure 4.26: Light Level Sensing Circuit

In this basic darkness detection circuit, the LDR light dependent resistor and VR1 potentiometer form an adjustable arm of a network of single resistance bridges, also commonly known as the Wheatstone bridge, while the two fixed resistors R_1 and R_2 form the other arm. Both sides of the bridge form potential dividing networks through the supply voltage whose outputs V_1 and V_2 are connected to the non-inverting and inverting voltage inputs respectively of the op amp. The op amp is configured as a differential amplifier also known as a feedback voltage comparator whose output voltage condition is determined by the difference between the two input signals or voltages, V_1 and V_2. The combination of resistors R_1 and R_2 forms a fixed voltage reference at input V_2, established by the ratio of the two resistors. The LDR - VR_1 combination provides a variable voltage input V_1 proportional to the light level detected by the photoresistor. As with the above circuit, the output of the op amp is used to control a relay, which is protected by a freewheeling diode, D_1. When the light level detected by the LDR and its output voltage falls below the reference voltage set in V_2, the output of the op amp changes state by activating the relay and changing the connected load. Similarly, as the light level increases, the output will change again and turn off the relay. The hysteresis of the two switching points is established by the feedback resistor R_f that can be chosen to provide any suitable amplifier voltage gain. The operation of this type of light sensor circuit can also be reversed to turn on the relay when the light level exceeds the reference voltage level and vice versa by reversing the positions of the LDR light sensor and VR_1 potentiometer. The potentiometer can be used to "preset" the switching point of the differential amplifier to any particular light level, making it ideal as a simple light sensor project circuit.

4.7.2 Photo Junction Devices

Photo junction devices are basically PN junction light sensors or detectors made of PN silicon semiconductor junctions that are light sensitive and can detect both visible light and infrared light levels. Photojunction devices are specifically designed to detect light, and this class of photoelectric light sensors includes the photodiode and the phototransistor.

The Photodiode

Figure 4.27: Photo-diode

The construction of the photodiode light sensor is similar to that of a conventional PN junction diode, except that the outer casing of the diodes is transparent or has a transparent lens to focus light on the PN junction for increased sensitivity. The junction will respond to light, especially at longer wavelengths, such as red and infrared, rather than visible light. This feature may be a problem for diodes with clear or glass bead bodies, such as the 1N4148 signal diode. LEDs can also be used as photodiodes as they can emit and detect light from their junction. All PN junctions are light sensitive and can be used in a photoconductive unbiased voltage mode with the photodiode PN junction always "reverse biased" so that only diode leakage or dark current can flow. I / V curves) of a photodiode with no light at its junction (dark mode) is very similar to a normal signal or rectifier diode. When the photodiode is forward biased, there is an exponential increase in current, the same as for a normal diode. When reverse bias is applied, a small reverse saturation current appears causing an increase in the depletion region, which is the sensitive part of the junction. Photodiodes can also be connected in a current mode using a fixed bias voltage across the junction. The current mode is very linear over a wide range.

Construction and characteristics of the photodiode

When used as a light sensor, a dark photodiode current (0 lux) is approximately 10uA for geranium and 1uA for silicon diodes. When light falls on the junction, more hole / electron pairs are formed and the leakage current increases. This leakage current increases as the illumination of the junction increases. Therefore, the photodiode current is directly proportional to the intensity of the light falling on the PN junction. One main advantage of photodiodes when used as light sensors is their fast response to changes in the light levels, but one disadvantage of this type of photo device is the relatively small current flow even when fully lit. Figure 4.28 shows a photo-current-to-voltage converter circuit using an operational amplifier as the amplifying device. The output voltage (V_{out}) is given as $V_{out} = I_p \times R_f$ and which is proportional to the light intensity characteristics of the photodiode. This type of circuit also utilizes the characteristics of an operational amplifier with two input terminals at about zero voltage to operate the photodiode without bias. This zero bias op amp configuration provides a high impedance load to the photodiode, resulting in less influence from dark current and a wider linear range of photocurrent relative to the intensity of radiant light. The Cf capacitor is used to avoid oscillation or peak gain and to set the output bandwidth ($1 / 2\pi RC$).

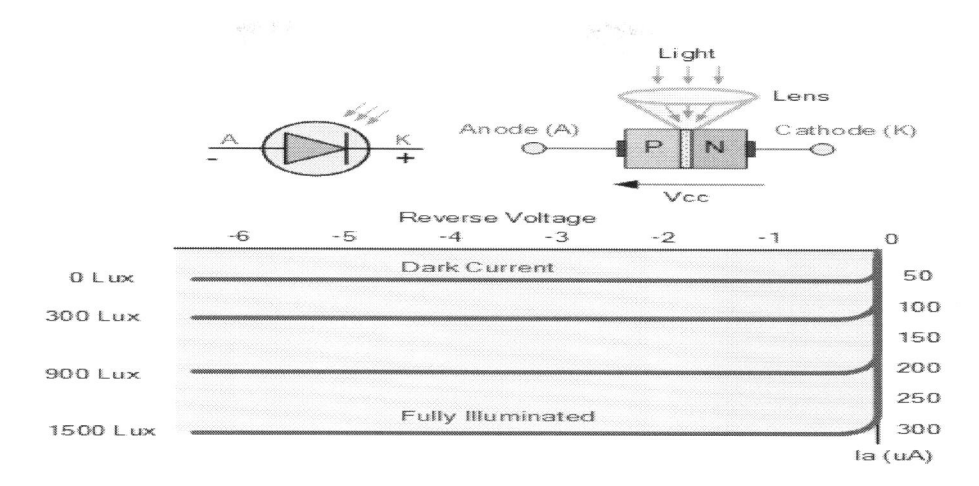

Figure 4.28: Characteristics of Photodiode

Photo-diode Amplifier Circuit

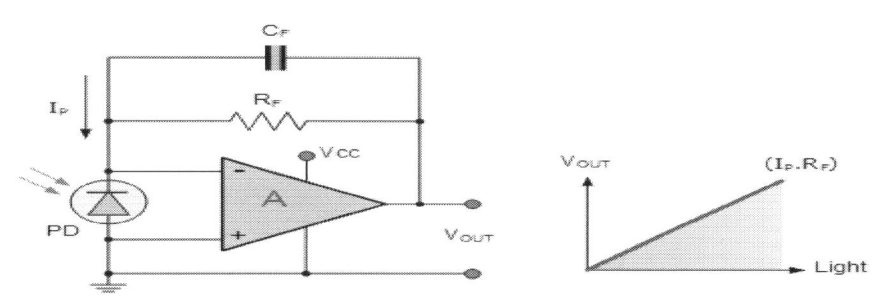

Figure 4.29: Photo-diode Amplifier Circuit

Photodiodes are highly versatile light sensors that can turn your current flow on and off in nanoseconds and are commonly used in cameras, light meters, CD and DVD-ROM drives, TV remote controls, scanners, fax machines, and copiers , etc. ., and when integrated into operational amplifier circuits such as infrared spectrum detectors for fiber optic communications, burglar alarm motion detection circuits and numerous imaging, laser scanning and positioning systems, etc.

4.7.3 The Phototransistor

An alternative photographic junction device to the photodiode is the phototransistor, which is basically an amplified photodiode. The phototransistor light sensor has a reverse biased collector base PN junction that exposes it to the radiant light source. Phototransistors work in the same way as the photodiode, except that they can provide current gain and are much more sensitive than the photodiode with currents that are 50 to 100 times greater than those of the standard photodiode and any normal transistor can easily be converted into a sensor. of phototransistor light by connecting a photodiode between the collector and the base. Phototransistors consist primarily of an NPN bipolar transistor with its large base region electrically disconnected, although some phototransistors allow a base connection to control sensitivity, and which uses light photons to generate a base current which in turn causes a collector to emit current. to issue. Most phototransistors are NPN type whose outer casing is transparent or has a transparent lens to focus light at the base junction to increase sensitivity.

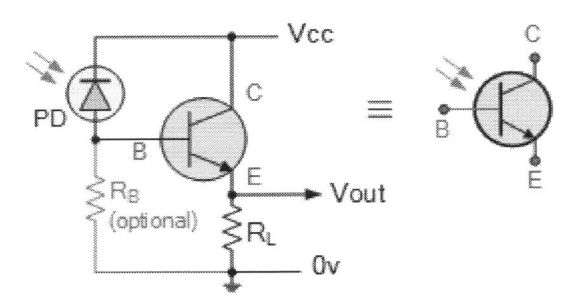

Figure 4.30: Photo-Transistor

Photo-Transistor Construction and Characteristics

In the NPN transistor, the collector is positively polarized with respect to the emitter, so that the base / collector junction is reverse polarized. Therefore, with no light at the junction, normal leakage or dark current flows are very small. When the light falls on the base, more pairs of electrons / holes are formed in this region and the transistor amplifies the current produced by this action.

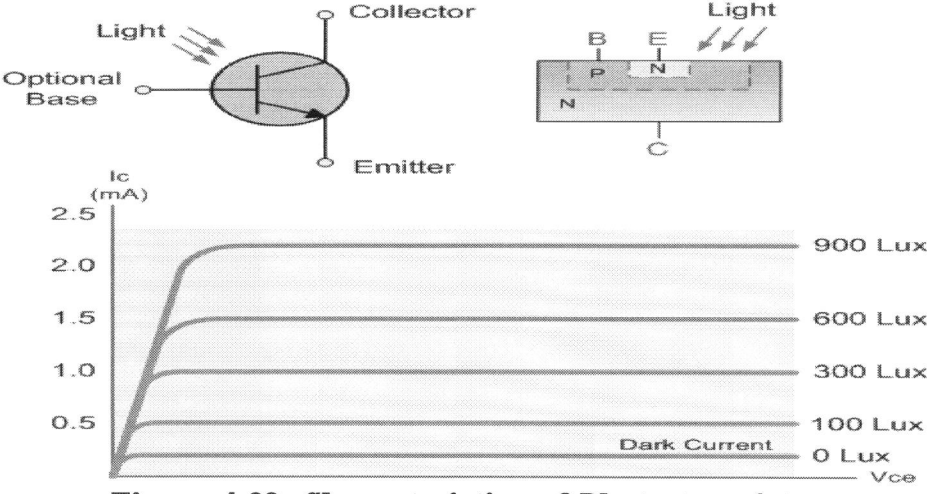

Figure 4.32: Characteristics of Photo-transistor

Generally, the sensitivity of a phototransistor is a function of the DC current gain of the transistor. Therefore, the general sensitivity is a function of the collector current and can be controlled by connecting a resistor between the base and the emitter, but for very high sensitivity optocoupler type applications Darlington phototransistors are generally used.

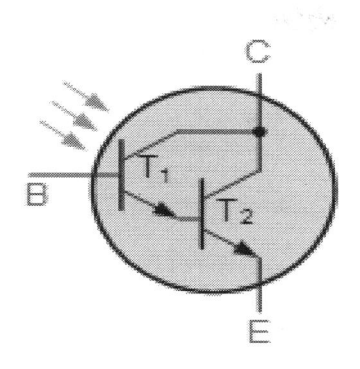

Figure 4.33: Photo-Darlington

Photo Darlington transistors use a second NPN bipolar transistor to provide additional amplification or when higher sensitivity is required from a photodetector due to low light levels or selective sensitivity, but its response is slower than that of a common NPN phototransistor. Photo Darlington devices consist of a normal phototransistor whose emitter output is coupled to the base of a larger NPN bipolar transistor. Because the configuration of a Darlington transistor provides a current gain equal to a product of the current gains of two individual transistors, a Darlington photographic device produces a very sensitive detector.

Typical applications for phototransistor light sensors are in optoisolators, slotted opto-switches, light beam sensors, fiber optics, and TV-type remote controls, etc. Infrared filters are sometimes required to detect visible light. Another type of photo junction semiconductor light sensor worth mentioning is the photo-thyristor. This is a light activated thyristor or silicon controlled rectifier, SCR that can be used as a light activated switch in AC applications. However, its sensitivity is usually very low compared to equivalent photodiodes or phototransistors. To help increase their sensitivity to light, photo-thyristors are made thinner around the door junction. The disadvantage of this process is that it limits the amount of anode current that can change. Then, for higher AC applications, they are used as pilot devices in optocouplers to change larger, more conventional thyristors.

4.7.4 Photovoltaic Cells

The most common type of photovoltaic light sensor is the solar cell. Solar cells convert light energy directly into DC electrical energy in the form of voltage or current at resistive charging power, such as a light, battery, or motor. So photovoltaic cells are similar in many ways to a battery in that they supply direct current. However, unlike the other photographic devices we've seen previously, which use the intensity of light even from a torch to operate; photovoltaic solar cells work best using radiant energy from the sun. Solar cells are used in many different types of applications to offer an alternative energy source from conventional batteries, such as in calculators, satellites and now in homes that offer a form of renewable energy.

4.8 Temperature Sensors

There are two main types of bimetallic strips based primarily on their movement when subject to changes in temperature. There are "fast acting" types that produce instantaneous "ON / OFF" or "OFF / ON" action on electrical contacts at a set temperature point, and slower "slow acting" types that gradually change its position as the temperature changes. Quick acting thermostats are commonly used in our homes to control the temperature set point of ovens, irons, immersion hot water tanks and can also be found on the walls to control the home heating system. Creeper types generally consist of a spiral or bimetallic coil that slowly unwinds or coils as the temperature changes. In general, creeper type bimetallic strips are more sensitive to temperature changes than standard pressure ON / OFF types as the strip is longer and thinner, making them ideal for use in temperature gauges and dials etc. Although they are very cheap and available in a wide range. The operating range, a main disadvantage of standard fast acting type thermostats when used as a temperature sensor, is that they have a large hysteresis range from the time the contacts are opened. until they close again. For example, it can be set to 20oC but it cannot be opened up to 22oC or closed again up to 18oC. Therefore, the temperature swing range can be quite high. Commercially available bimetal thermostats for home use have temperature adjusting screws that allow for preest

4.8.1 The Thermistor

The thermistor is another type of temperature sensor; whose name is a combination of the words THERM-ally sensible res-ISTOR. A thermistor is a special type of resistance that changes its physical resistance when exposed to temperature changes.

Figure 4.34: Thermistor

Thermistors are generally made of ceramic materials such as glass-coated nickel, manganese, or cobalt oxides, which easily damages them. Its main advantage over fast acting types is its speed of response to any change in temperature, precision and repeatability. Most thermistor types have a negative temperature resistance coefficient (NTC), that is, their resistance value goes down with an increase in temperature and of course there are some that have a positive temperature coefficient (PTC) . the resistance value increases with an increase in temperature. Thermistors are built from a ceramic-type semiconductor material that uses metal oxide technology such as manganese, cobalt, and nickel, etc. The semiconductor material is generally formed into small pressed discs or balls that are hermetically sealed to give a relatively quick response to any change in temperature. . Thermistors are classified by their resistive value at room temperature (generally at 25 ° C), their time constant (the reaction time to temperature change), and their power rating with respect to the current flowing through them. Like resistors, thermistors are available with resistance values at room temperature from 10 MΩ to just a few ohms, but those types with kilo ohm values are generally used for detection purposes. Thermistors are passive resistive devices, which mean that we need to pass a current through it to produce a measurable voltage output. The thermistors are then generally connected in series with a suitable bias resistor to form a potential dividing network and the choice of resistance provides a voltage output at some predetermined point or temperature value, for example:

4.8.2 Thermistor Temperature Sensors

The following thermistor has a resistance value of 10KΩ at 25oC and a resistance value of 100Ω at 100oC. Calculate the voltage drop across the thermistor, and therefore its output voltage (V_{out}) for both temperatures when connected in series with a 1kΩ resistor through a 12v power supply.

At 25°C

$$Vout = \frac{R_2}{R_1+R_2} xV = \frac{1000}{10000+1000} x12v = 1.09v \quad (4.8)$$

At 100°C

$$Vout = \frac{R_2}{R_1+R_2} xV = \frac{1000}{100+1000} x12v = 10.9v \quad (4.9)$$

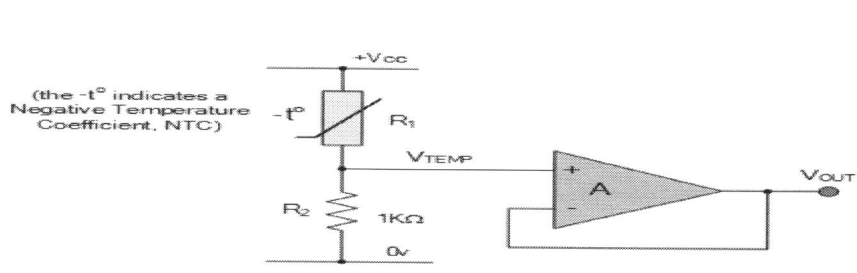

Figure 4.35: Temperature Sensors Using Thermistor

By changing the value of the fixed resistance of R2 (in our example 1kΩ) to a potentiometer or preset, a voltage output can be obtained at a predetermined temperature set point, for example, 5v output at 60oC and by varying the potentiometer, a particular output voltage level can be obtained over a wider temperature range. However, it should be noted that thermistors are nonlinear devices and their standard resistance values at room temperature are different between different thermistors, which is mainly due to the semiconductor materials from which they are made. The thermistor has an exponential change with temperature, and therefore has a temperature constant Beta (β) that can be used to calculate its resistance for any given temperature point. However, when used with a series resistor, such as in a voltage divider network or a Whetstone bridge arrangement, the current obtained in response to a voltage applied to the divider / bridge network is linear with temperature. . Then the output voltage across the resistor becomes linear with the temperature.

4.8.3 Resistive Temperature Detectors (RTD)

Another type of electrical resistance temperature sensor is the Resistance Temperature Detector or RTD. RTDs are precision temperature sensors made of high purity conductive metals such as platinum, copper or nickel wound on a coil and whose electrical resistance changes as a function of temperature, similar to that of the thermistor. Thin film RTDs are also available. These devices have a thin film of platinum paste that is deposited on a white ceramic substrate.

Figure 4.36: A Resistive RTD

Resistive temperature detectors have positive temperature coefficients (PTC), but unlike the thermistor, their output is extremely linear and produces very accurate temperature measurements. However, they have a very low thermal sensitivity, that is, a temperature change only produces a very small output change, for example, 1Ω / oC. The most common types of RTDs are made of platinum and are called a resistance thermometer. The equation below relates the change in resistance for a small change in temperature.

$$R = R_O(1 + \alpha \Delta T) \quad (4.10)$$

R = resistance at new temperature
R_0 = resistance at reference temperature (0°C)
α = temperature coefficient of wire (0.00385/°C for pt)
ΔT = change in temperature

Like the thermistor, RTDs are passive resistive devices and by passing a constant current through the temperature sensor it is possible to obtain an output voltage that increases linearly with temperature. A typical RTD has a base resistance of approximately 100Ω at 0ºC, increasing to approximately 140Ω at 100ºC with an operating temperature range of -200 to + 600ºC. Because the RTD is a resistive device, we need to pass a current through them and monitor the resulting voltage. However, any variation in resistance due to self-heating of the resistive cables as current flows through it, I^2R, (Ohm's Law) causes an error in the readings. To avoid this, the RTD is generally connected to a Whetstone Bridge network that has additional lead wires for lead compensation and / or connection to a constant current source.

4.8.4 The Thermocouple

The thermocouple is by far the most widely used type of all types of temperature sensors. Thermocouples are popular because of their simplicity, ease of use, and speed of response to temperature changes, primarily due to their small size. Thermocouples also have the widest temperature range of all temperature sensors from below -200 ° C to over 2000 ° C. Thermocouples are thermoelectric sensors that basically consist of two different metal junctions, such as copper and constantan, which are welded or set together. One junction is held at a constant temperature called the reference junction (cold), while the other is the measurement junction (hot). When the two junctions are at different temperatures, a voltage develops across the junction that is used to measure the temperature sensor as shown below.

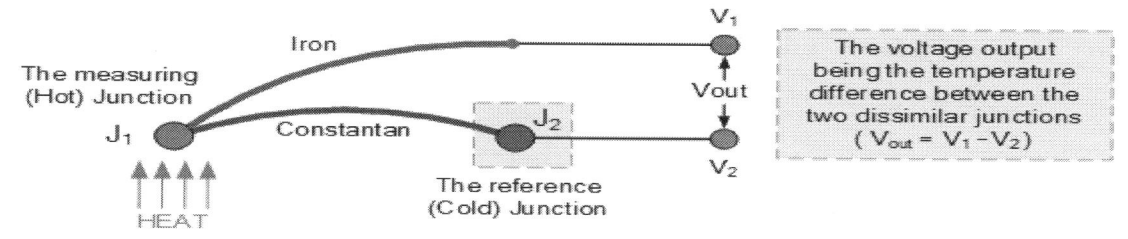

Figure 4.37: Thermocouple Construction

The operating principle of a thermocouple is very simple and basic. When fused, the union of the two different metals, such as copper and constantane, produces a "thermoelectric" effect that provides a constant potential difference of only a few millivolts (mV) between them. The difference in voltage between the two junctions is called the "Seebeck effect," since a temperature gradient is generated along the lead wires that produce an emf. So, the output voltage of a thermocouple is a function of temperature changes. If both junctions are at the same temperature, the potential difference between the two junctions is zero, in other words, there is no voltage output like $V_1 = V_2$. However, when the junctions are connected within a circuit and both are at different temperatures, a voltage output will be detected in relation to the temperature difference between the two junctions, $V_1 - V_2$. This voltage difference will increase with temperature until the maximum voltage level of the junction is reached and this is determined by the characteristics of the two different metals used. Thermocouples can be made of a variety of different materials that allow extreme temperatures to be measured from -200 ° C to over + 2000 ° C. With such a wide selection of materials and temperature range, internationally recognized standards have been developed complete with Thermocouple color codes to allow the user to choose the correct thermocouple sensor for a particular application. The British color code for standard thermocouples is provided below.

The three most common thermocouple materials previously used for general temperature measurement are iron-constantane (type J), copper-constantane (type T), and nickel-chromium (type K). The output voltage of a thermocouple is very small, only a few millivolts (mV) for a 10°C change in temperature difference, and due to this small voltage output some form of amplification is generally required.

Table 4.3: British colour code for standard thermocouples

Type	Metals	Coefficient (µV/°C)	Operating temp range (°C)
E	Chromel-	62	-265 to 1000
J	Iron-constantan	51	-205 to 750
K	Chromel-alumel	40	-270 to 1365
S	Platinum-rhodium	7	0 to 1750
T	Copper-constantan	40	-265 to 400

The Seebeck coefficients in the table above can be used in the following equation to find the voltage at the measurement end of the sensor.

$$V = s (\Delta T) \qquad (4.12)$$

V is in volts, s is the Seebeck coefficient, ΔT is the difference in temperature between the two ends

Thermocouple Amplification

The type of amplifier, either discrete or in the form of an op-amp, must be carefully selected as good drift stability is required to avoid recalibration of the thermocouple at frequent intervals. This makes the type of amplifier and instrumentation preferable for most applications. Other types of temperature sensors not mentioned here include, semiconductor junction sensors, infrared and thermal radiation sensors, medical-type thermometers, indicators, and color-changing inks or dyes.

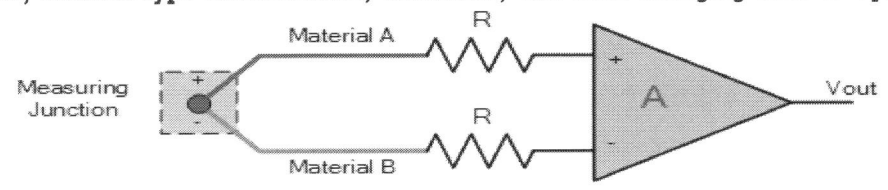

Figure 4.38: Thermocouple Amplification Circuit

4.9 Position Sensors

Position sensors detect the position of something, which means they are referenced to or from some fixed point or position. These types of sensors provide "positional" feedback. One method of determining a position is to use "distance", which could be the distance between two points, such as the distance traveled or away from some fixed point, or by "rotation" (angular movement). For example, the rotation of a robot's wheel to determine the distance traveled by the ground. Either way, position sensors can detect the movement of an object in a straight line using linear sensors or by its angular movement using rotational sensors.

4.9.1 The Potentiometer

The most widely used of all "Position Sensors" is the potentiometer because it is an inexpensive and easy to use position sensor. It has a wiper contact connected to a mechanical axis that can be angular (rotational) or linear (sliding type) in its movement, and that causes the resistance value between the wiper / slider and the two end connections to change giving an electrical signal output that has a proportional relationship between the actual position of the cleaner on the resistive track and its resistance value. In other words, resistance is proportional to position.

Figure 4.39: Potentiometer

Potentiometers come in a wide range of designs and sizes, such as the commonly available round rotation type or the longer, flatter linear sliders. When used as a position sensor, the moving object connects directly to the rotary axis or the potentiometer slider. A DC reference voltage is applied across the two external fixed connections that form the resistive element. The output voltage signal is taken from the terminal of the sliding contact cleaner. This configuration produces a voltage or potential divider type circuit output that is proportional to the position of the axis. So, for example, if you apply a voltage of, say, 10v across the resistive element of the potentiometer, the maximum output voltage would be equal to the supply voltage at 10 volts, with a minimum output voltage equal to 0 volts. The potentiometer cleaner will then vary the output signal from 0 to 10 volts, with 5 volts indicating that the cleaner or slider is in its middle or center position.

Potentiometer Construction

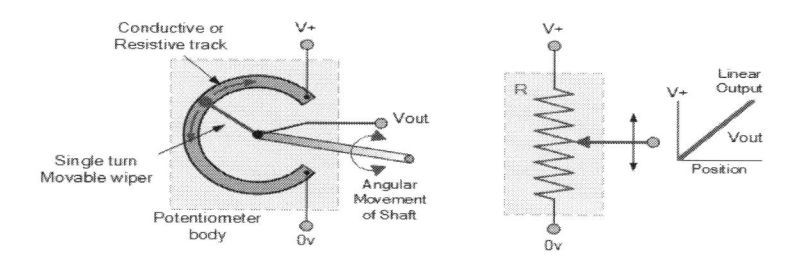

Figure 4.40: Construction of Potentiometer

The potentiometer's output signal (V_{out}) is taken from the center wiper connection as it moves along the resistive track, and is proportional to the angular position of the shaft.

Circuit Diagram of Positional Sensing

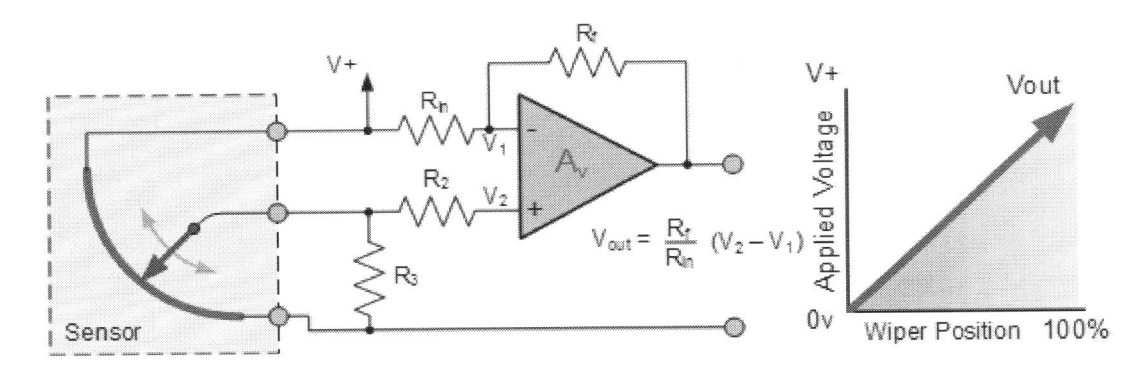

Figure 4.41: Circuit Diagram of Positional Sensing

Although the resistive potentiometer position sensors have many advantages: low cost, low technology, easy to use, etc., as a position sensor they also have many disadvantages: wear due to moving parts, low precision, low repeatability and frequency response limited. But there is a major disadvantage of using the potentiometer as a position sensor. The range of motion of your wiper or slider (and therefore the output signal obtained) is limited to the physical size of the potentiometer being used. For example, a single turn rotary potentiometer generally only has a fixed mechanical rotation of between 0o and approximately 240 to 330o maximum. However, multi-turn containers of up to 3600o (10 x 360o) mechanical rotation are also available. Most potentiometers use carbon film for their resistive track, but these guys are electrically noisy (the crunch on a radio volume control), and they also have a short mechanical life. Wire wound pots, also known as rheostats, can also be used in the form of straight wire or coiled resistive wire, but rolled pots suffer from resolution problems as their cleaner jumps from one wire segment to the next producing a logarithmic (LOG) that produces errors in the output signal. These also suffer from electrical noise. For high-precision, low-noise applications, conductive plastic resistive element type or cermet type polymer film potentiometers are now available. These pots have a low friction electrically linear resistive (LIN) track that provides low noise, long life and excellent resolution, and are available as multi-turn and single-turn devices. Typical applications for this type of high-precision position sensor are found in computer game joysticks, steering wheels, industrial and robot applications.

4.9.2 Inductive Position Sensors
4.9.2.1 Linear Variable Differential Transformer

One type of positional sensor that does not suffer from mechanical wear problems is the "Linear Variable Differential Transformer" or LVDT for short. This is an inductive type position sensor that works on the same principle as the AC transformer that is used to measure motion. It is a very precise device for measuring linear displacement and whose output is proportional to the position of its moving core. It basically consists of three coils wound in a hollow tube former, one forming the primary coil and the other two coils forming identical connected secondary connections. Electrically together in series but 180o out of phase on either side of the primary coil. A moving soft iron ferromagnetic core (sometimes called"armor") that is attached to the object being measured slides or moves up and down within the tubular body of the LVDT. A small AC reference voltage called an "excitation signal" (2-20V rms, 2-20kHz) is applied to the primary winding which in turn induces an EMF signal in the two adjacent secondary windings (transformer principles).

If the soft iron magnetic core armature is exactly in the center of the tube and the windings, "null position", the two emf induced in the two secondary windings cancel each other as they are 180o out of phase, so the resulting output voltage is zero As the core shifts slightly to one side or the other from this zero or zero position, the induced voltage on one of the secondary will be higher than that of the other secondary and an output will occur. The polarity of the output signal depends on the direction and displacement of the moving core. The greater the movement of the soft iron core from its central null position, the greater the resulting output signal. The result is a differential voltage output that varies linearly with the position of the cores. Therefore, the output signal of this type of position sensor has both amplitudes, which is a linear function of the nuclei displacement and a polarity indicating the direction of movement. The phase of the output signal can be compared to the excitation phase of the primary coil, allowing suitable electronic circuits such as the LVDT AD592 sensor amplifier to know which half of the coil the magnetic core is in and therefore know the direction of travel.

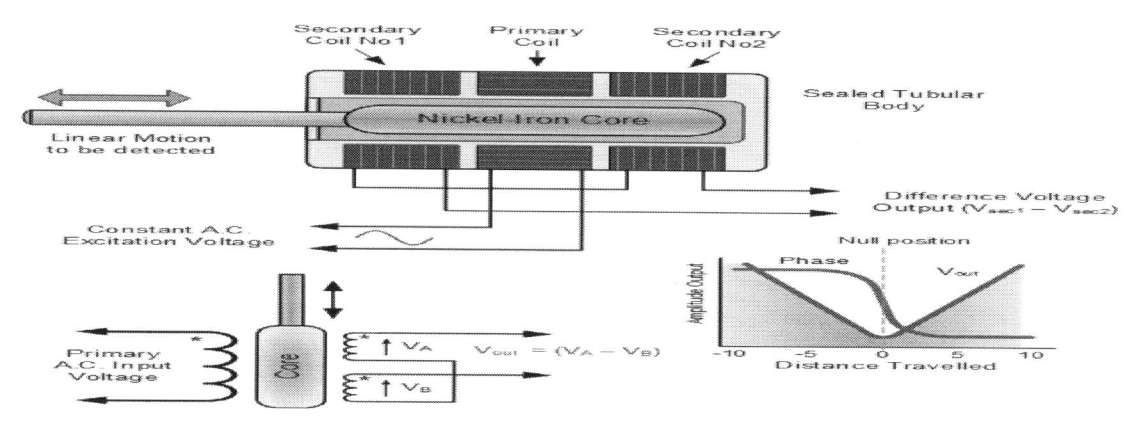

Figure 4.42: Linear Variable Differential Transformer

When the armature moves from one end to the other through the center position, the output voltages change from maximum to zero and back to maximum, but in the process its phase angle changes by 180 degrees. This allows the LVDT to produce an AC output signal whose magnitude represents the amount of motion from the center position and whose phase angle represents the direction of core motion. A typical application for a linear variable differential transformer (LVDT) sensor would be as a pressure transducer, if the measured pressures push against a diaphragm to produce a force. The force is converted into a sensor-readable voltage signal. The advantages of the linear variable differential transformer or LVDT compared to a resistive potentiometer are that its linearity, which is its displacement voltage output, is excellent, very good precision, good resolution, high sensitivity and frictionless operation. They are also sealed for use in harsh environments.

4.9.2.2 Inductive Proximity Sensors or Presence Sensor

A proximity sensor is a sensor capable of detecting the presence of nearby objects without any physical contact. A proximity sensor often emits an electromagnetic or electrostatic field, or a beam of electromagnetic radiation (infrared, for example), and looks for changes in the field or the return signal. The object that is detected is often referred to as the proximity sensor target. Different proximity sensor targets require different sensors. For example, a capacitive or photoelectric sensor might be suitable for a plastic target, while an inductive proximity sensor requires a metal target. Proximity sensors can have high reliability and a long functional life due to the absence of mechanical parts and the lack of physical contact between the sensor and the detected object. The maximum distance this sensor can detect is defined as its "nominal range". Some sensors have nominal range adjustments or means to report a graduated detection distance. Proximity sensors are also used in machine vibration monitoring to measure variation in the distance between a shaft and its support bearing. This is common in large steam turbines, compressors, and engines that use sleeve bearings.

Some proximity sensors are divided into two halves and if the two halves are separated, a signal is activated. An example of a proximity sensor application is for window security. When the window opens, an alarm is triggered. Another type of inductive position sensor in common use is the inductive proximity sensor, also called the Eddy current sensor. While they don't actually measure angular displacement or rotation, they are primarily used to detect the presence of an object in front of or very close to it, hence its name "Proximity Sensors". Proximity sensors are non-contact position sensors that use a magnetic field for detection, with the simplest magnetic sensor being the reed switch. In an inductive sensor, a coil is wound around an iron core within an electromagnetic field to form an inductive loop. When a ferromagnetic material is placed within the eddy current field generated around the inductive sensor, such as a ferromagnetic metal plate or metal screw, the inductance of the coil changes significantly. The proximity sensor detection circuit detects this change by producing an output voltage. Therefore, inductive proximity sensors operate under the electrical principle of Faraday's Law of Inductance.

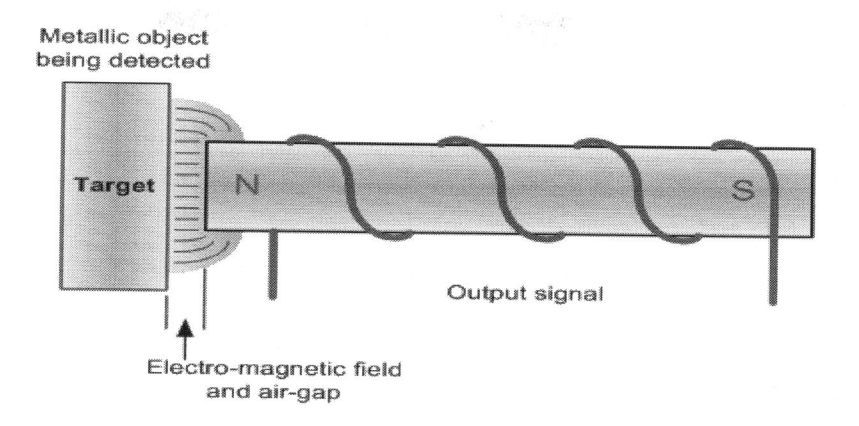

Figure 4.43: Inductive Proximity Sensors

An inductive proximity sensor has four main components; The oscillator that produces the electromagnetic field, the coil that generates the magnetic field, the detection circuit that detects any change in the field when an object enters, and the output circuit that produces the output signal, either with normally closed (NC) or normally open contacts (NO). Inductive proximity sensors allow the detection of metallic objects in front of the sensor head without any physical contact of the object itself being detected. This makes them ideal for use in dirty or wet environments. The "detection" range of proximity sensors is very small, typically 0.1mm to 12mm. In addition to industrial applications, inductive proximity sensors are also commonly used to control traffic flow by changing traffic lights at crossings and road junctions. Rectangular inductive wire loops are buried in the asphalt road surface. When a car or other road vehicle passes over this inductive loop, the vehicle's metal body changes the inductance of the loops and activates the sensor alerting the traffic light controller that a vehicle is waiting.

Figure 4.44: Proximity Sensor

A main disadvantage of this type of position sensors is that they are "omnidirectional", that is, they will detect a metallic object above, below or to the side. Furthermore, they do not detect non-metallic objects, although capacitive proximity sensors and ultrasonic proximity sensors are available. Other commonly available magnetic position sensors include: reed switches, Hall effect sensors, and variable reluctance sensors.

4.9.2.3 Rotary Encoders

Rotary encoders are another type of position sensor that resembles the potentiometers mentioned above, but they are non-contact optical devices used to convert the angular position of a rotary axis into an analog or digital data code. In other words, they convert mechanical motion into an electrical signal (preferably digital). All optical encoders operate on the same basic principle. Light from an LED or an infrared light source passes through a high resolution spinning encoded disk that contains the necessary code patterns, be it binary, gray code, or BCD. Photodetectors scan the disk as it spins, and an electronic circuit processes the information digitally as a stream of binary output pulses that are fed to counters or controllers that determine the actual angular position of the axis. There are two basic types of rotary optical encoders, incremental encoders and absolute position encoders.

1. Incremental Encoder

Incremental encoders, also known as quadrature encoders or relative rotary encoders, are the simplest position sensors. Its output is a series of square wave pulses generated by a photocell arrangement when the encoded disc, with evenly spaced dark and transparent lines called segments on its surface, moves or rotates beyond the light source.

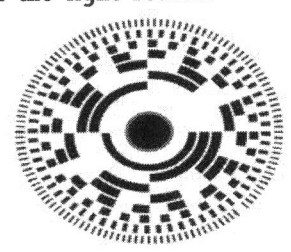

Figure 4.45: Encoder Disk

The encoder produces a square wave pulse current which, when counted, indicates the angular position of the rotary axis. Incremental encoders have two separate outputs called "quadrature outputs". These two outputs move at 90° out of phase from each other with the axis rotation direction determined from the output sequence. The number of transparent and dark segments or slots on the disk determines the resolution of the device, and increasing the number of lines in the pattern increases the resolution per degree of rotation. Typical encoded discs have a resolution of up to 256 pulses or 8 bits per rotation. The simplest incremental encoder is called a tachometer. It has a single square wave output and is often used in unidirectional applications where only basic position or speed information is required. The "quadrature" or "sine wave" encoder is the most common and has two square output waves commonly called channel A and channel B. This device uses two photographic detectors, slightly offset from each other by 90o, producing two separate outputs sine and cosine signals

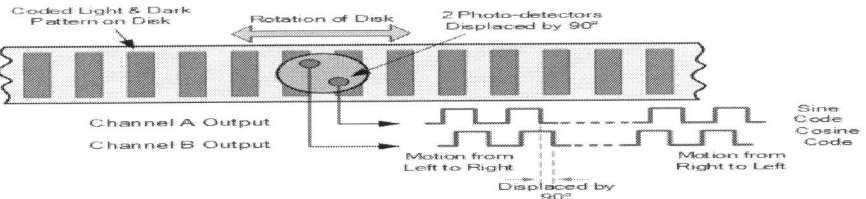

Figure 4.46: Incremental Encoder

Using the Arc Tangent math function, the angle of the axis can be calculated in radians. In general, the optical disk used in rotary position encoders is circular, then the resolution of the output will be given as: $\theta = 360 / n$, where n is equal to the number of segments in the encoded disk. So, for example, the number of segments required to give an incremental encoder a resolution of 1º will be: $1º = 360 / n$, therefore n = 360 windows, etc. The direction of rotation is also determined by looking at which channel produces an output first, either channel A or channel B giving two directions of rotation, A leads B or B leads A. This arrangement is shown below.

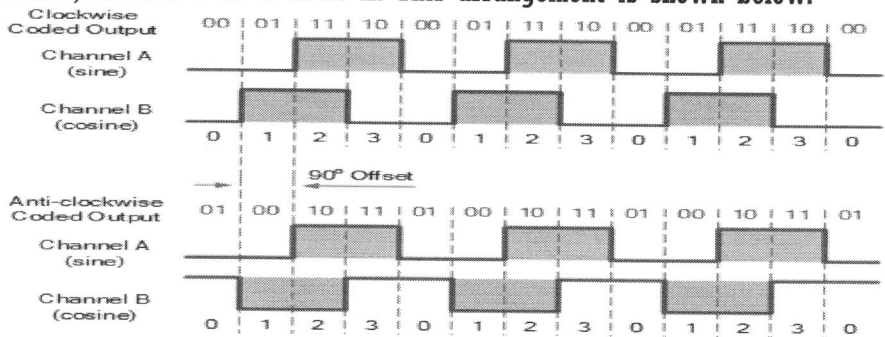

Figure 4.47: Incremental Encoder Output

A major disadvantage of incremental encoders when used as position sensors is that they require external counters to determine the absolute angle of the axis within a given rotation. If the power is turned off momentarily, or if the encoder loses a pulse due to noise or a dirty disc, the resulting angular information will fail. One way to overcome this disadvantage is to use absolute position encoders.

2. Absolute Position Encoder

Absolute position encoders are more complex than quadrature encoders. They provide a unique exit code for each rotation position that indicates both position and direction. His coded disc consists of multiple concentric "tracks" of light and dark segments. Each track is independent with its own photo detector to simultaneously read a unique encoded position value for each angle of movement. The number of tracks on the disk corresponds to the encoder's "bit" binary resolution, so a 12-bit absolute encoder would have 12 tracks and the same encoded value only appears once per revolution.

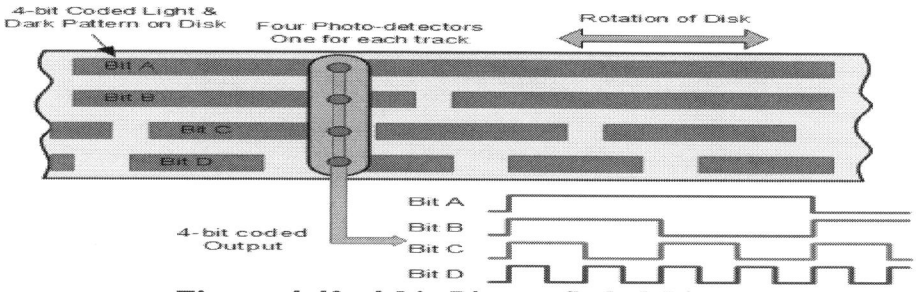

Figure 4.48: 4-bit Binary Coded Disc

A major advantage of an absolute encoder is its non-volatile memory that retains the exact encoder position without the need to return to an "initial" position if power fails. Most rotary encoders are defined as "single-turn" devices, but absolute multi-turn devices are available, which get feedback over several revolutions by adding additional code discs. if the absolute position of the read / write heads of the drives or printers / plotters is controlled to precisely position the print heads on the paper.

4.10 Force Sensors

Force sensors are used to obtain an accurate determination of tensile and / or pressure forces. The force sensor creates an electrical signal that corresponds to the force measurement to be used for further evaluation or process control. Force sensors are commonly used in motor vehicle assemblies, such as brakes, suspension units, and airbag systems. A force sensor generally measures the applied force from the proportional deformation of a spring element; the greater the force, the more this element is deformed. Many force sensors employ the piezoelectric principle exhibited by quartz. Under load, quartz crystals produce an electric charge proportional to the applied mechanical load; the greater the load, the greater the load. Therefore, in piezoelectric force sensors, quartz serves as a spring element and as a measurement transducer. A deformation gauge is a device used to measure the deformation of an object. The strain gauge is the fundamental sensor element for many types of sensors, including pressure sensors, load cells, torque sensors, and position sensors. Most strain gages are made of aluminum. They consist of a resistive sheet pattern that is mounted on a backing material and when the sheet is stressed, the sheet resistance changes. This results in a signal output, related to the voltage value. Deformation is defined as the deformation caused by the action of stress (that is, the force per unit area in a given plane) on a body. Tension is the change in shape (specifically length) and / or size caused by stress.

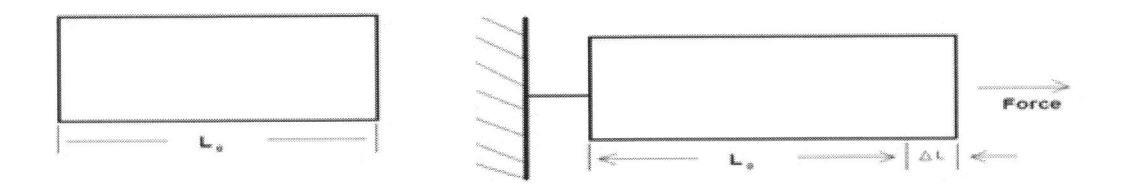

Figure 4.49: Force Sensors

$$Strain = \frac{\Delta L}{L_0} \qquad Stress = \frac{F}{A} \tag{4.13}$$

Where ΔL = change of length, L_0 = Original length, F = Total force applied to gage, A = Cross sectional area of gage

Solid metal bars do not stretch very much, but all metals are somewhat elastic as long as the force has not exceeded the elastic limit of the metal. That is, a piece of metal stretches when a force is applied, but returns to its original dimensions when the force is removed. If the metal object is subjected to a force greater than its elastic limit, the metal object will deform and will not return to its original dimensions when the force is removed. The stress to strain ratio is a constant value for materials such as steel, brass and / or aluminum and the dimensions of the material are not important. This stress-strain relationship, called Young's modulus, is recalculated and listed in the engineering reference materials. The following equation finds the resistance of a strain gage if its original resistance, length change and gauge factor are known.

$$R = R_O(1 + \frac{\Delta L}{L_O} x GF) \qquad (4.14)$$

Where R = gage resistance under stress, R_0= original resistance of gage, ΔL = change of length, L_0= original length, GF = gage factor.

4.11 Motion Sensors

Motion sensors are designed to measure the rate of change in position, location, or displacement of an object that is occurring. If the position of an object is changing as a function of time, the first derivative gives the velocity of the object, and if the velocity of the object is also changing, then the first derivative of velocity gives the acceleration. Displacement sensors measure the distance an object moves and can also be used to measure the height and width of the object. Optical and magnetic displacement sensors are used to detect the amount of linear displacement. Linear and angular displacement sensors are used for high-precision machining and measurement, to manufacture and test components with very tight dimensional tolerances. Magnetic displacement sensors consist of one or more magnets that produce an induction field and typically measure linear or rotational displacement that provides an output proportional to the absolute linear or rotary displacement of the position of the elements.

The simplest type of acceleration measures the movement of the mass by attaching the mass of the spring to the wiper arm of a potentiometer. In this way, the mass position is transmitted as a changing resistance. The linear variable differential transformer (LVDT) takes advantage of natural linear displacement to measure the displacement of mass. In these instruments, the LVDT core is the seismic mass. The displacements of the nucleus are converted directly into a linearly proportional voltage. These accelerometers generally have a natural frequency of less than 80 Hz and are commonly used for low-frequency, steady-state vibrations. Vibration sensors are sensors for measuring, displaying, and analyzing linear speed, displacement and proximity, or acceleration. They can be used independently or in conjunction with a data acquisition system. Vibration sensors are available in many forms. They can be raw detection elements, packaged transducers, or as a sensor or instrument system, incorporating features such as local or remote viewing and data recording. The main elements of importance in shock measurements are that the device has a natural frequency greater than 1 kHz and a range typically greater than 500 g. The primary accelerometer that can satisfy these requirements is the piezo type.

4.12 Fluid Sensors

A fluid pressure sensor detects a pressure difference between a detection pressure and a reference pressure and converts the difference into an electrical signal. Pressure sensors are used to measure gas or liquid pressures. Pressure measurements are generally made as absolute, gauge, or differential measurements. Absolute pressure sensors measure a pressure relative to a vacuum, measurement sensors measure a pressure relative to atmospheric pressure, and differential sensors measure a pressure difference between two inlets. Generally, a pressure sensor for detecting a gas or liquid pressure has a diaphragm that acts as a pressure sensing element and is configured such that the deflection (pressure deformation) of a diaphragm under fluid pressure applied to Through a pressure port it is converted into an electrical signal, allowing the fluid pressure to be measured.

A fluid flow sensor is a device to detect the speed of the fluid flow. Typically, a flow sensor is the sensing element used in a flow meter or data logging device to record the flow of fluids. There are several types of flow sensors and flow meters, including some that have a vane that is pushed by the fluid and can drive a rotary potentiometer or similar device. Other flow sensors are based on sensors that measure the heat transfer caused by the moving medium. The flow sensor can normally measure velocity, flow velocity, or total flow. Flow sensors are sometimes related to sensors called speed meters that measure the speed of fluids that flow through them. Flow sensor technology can be based on things like light, heat, electromagnetic properties, ultrasonic technologies, and many others on a broad spectrum. A flow sensor can operate by direct measurement or inferential measurement. Various types of flow sensors are non-mechanical and normally operate by the inferential method. A viscometer is an instrument used to measure the viscosity of a fluid. Viscometers only measure a flow condition. In general, either the fluid remains stationary and an object moves through it, or the object is stationary and the fluid passes through it. The drag caused by the relative movement of the fluid and a surface is a measure of the viscosity.

4.13 Environmental Sensors

Environmental sensors include those used to measure temperature, humidity, wind speed, barometric pressure, etc. They are often connected on a network. Temperature sensors are discussed in a preview. An analog humidity sensor measures air humidity relatively using a condenser-based system. The sensor is made of a film generally made of glass or ceramic. The insulating material that absorbs water is made of a polymer that absorbs and releases water depending on the relative humidity of the given area. This changes the charge level in the capacitor in the electrical circuit. A digital humidity sensor works through two micro sensors that are calibrated for the relative humidity of the given area. These are then converted to digital format with an analog to digital conversion. Most smoke detectors work by optical detection (photoelectric) or by physical process (ionization). They can also use both detection methods to increase sensitivity to smoke. In a photoelectric device, smoke can block a light beam by reducing the light that reaches a photocell used to trigger an alarm. Ionization detectors have an ionization chamber and a source of ionizing radiation. The released radiation particles ionize the oxygen and nitrogen atoms in the chamber. These positively charged atoms are attracted to the negative plate in the chamber and the electrons are attracted to the positive plate, generating a small continuous electric current. When the smoke enters the ionization chamber, the smoke particles adhere to the ions and neutralize them so that they do not reach the plate. The current drop between the plates activates the alarm.

Chapter -5: Generic Design Methodology and an IoT System Management

5.1 Technical Standards of IoT

An IoT project is the act of connecting any physical object to the Internet to collect and share data. Almost any physical object can be transformed into an IoT device if it can be connected to the Internet and controlled in that way. IoT hype is reaching high levels of data as sensor technologies like near-field communications (NFC) and radio frequency identification (RFID), wireless technologies, and other M2M tools become cheaper and more available. There are few technical standards for IoT, of which most of them are open standards, and the rest are standards organizations. The technical standards are:

- Auto-ID Labs - Auto Identification Center
- EPCglobal - Electronic Product code Technology
- FDA - U.S. Food and Drug Administration
- GS1
- IEEE - Institute of Electrical and Electronics Engineers
- IETF - Internet Engineering Task Force
- MTConnect Institute
- O-DF - Open Data Format
- O-MI - Open Messaging Interface
- OCF - Open Connectivity Foundation
- OMA - Open Mobile Alliance
- XSF - XMPP Standards Foundation

5.2 Generic Design Methodology

An IoT system refers to the integration of eight different important components: sensors / devices, connectivity, data processing, edge devices, gateway, cloud, analytics, and a user interface.

1) Sensors/Devices

Sensors or devices will collect data from your environment. This could be something like the temperature reading or even a full video. Depending on the purpose, sometimes even multiple sensors can be grouped together or can be part of one device. Your smartphone is a device that carries multiple sensors (camera, accelerometer, GPS, others). It can be a standalone or complete device, but the first step is to collect data from the environment.

2) Connectivity

Now, that data will be sent to the cloud. This will require connecting to the cloud on behalf of cell phones, satellites, Wi-Fi, Bluetooth, Low Power Wide Area Networks (LPWAN), or you can connect directly to the Internet using Ethernet. There is a tradeoff between power consumption, range, and bandwidth, so selecting the best connectivity option is subject to a specific IoT application, but they all do the same job of bringing data to the cloud.

3) Data Processing

After the data reaches the cloud, some software is used to do some kind of processing on it. This could be simple or complex, like using computer vision in video to recognize objects.

4) User Interface

Make the information useful to the end user in some way. This is possible through an alert to the user in the form of email, text, notification, etc. A user will have an interface that will allow him to proactively register in the system.

5) Edge Device

It is the device used to collect data from the field or actual physical location. The Edge device is basically a microcontroller with different sensors connected to it. Also known as a sensor node. The complexity of the device depends on the application to be developed. It can be a simple 8-bit microcontroller with bare metal programming or it can be a 32-bit microprocessor that uses some RTOS. The communications stack is an essential part for any edge device at IOT. This is the area of integrated systems development and requires knowledge of processor architecture, H / w interface, RTOS, and systems programming concepts.

6) Gateway

In large real-world IOT applications, there are many edge devices. Some kind of coordination between them is required. Here the gateway device comes into the picture. Typically, all edge devices are connected to the gateway by some means of communication. Gateway is smarter and more capable than edge devices and acts as an intermediary between edge devices and the cloud. The primary tasks for the gateway are device coordination, network administration, message routing, and more. It may be optional in some applications. Knowledge of network programming is useful in developing gateway software.

7) Cloud

IOT systems generate big data. This huge data must be stored somewhere. All data will be stored in the cloud. Edge devices generate a data stream, so the cloud must be able to store those data streams. Another task performed in the cloud is device management that includes device registration which includes device registry and device authentication. Knowledge of DBMS, data modelling, server administration are required for developing applications on cloud.

8) Analytics

There is no use of this large amount of data if we cannot get any idea of this. The analysis of this data provides us with information on the condition of the process or the monitored field. In real-time IOT applications, after analyzing the current status of the process, some type of trigger signal must be sent to the edge device to activate an actuator. Therefore, real-time analysis of the data flow is very important in IOT applications. Every time a new technology evolves, it redefines our existing way of life. With each part, we move toward greater comfort, a smarter and better life.

Here are some important predictions for IoT trends in 2021, as the Internet of Things is determined to benefit the business dimension.

Travel Business

Travel industries are interested in using IoT devices in their business operations and are using various strategies to present innovation to their customers. IoT devices will help these organizations collect and coordinate an immense amount of informational data from multiple sources to personalize and enhance the customer experience. It will encourage air travelers to find what they want and help industries offer more personalized services to their customers. Airlines and passengers alike will benefit from such amazing technology and connected people.

Manufacturing

Industrial Manufacturing is among the top three industries interested in embracing IoT and digital transformation. In the manufacturing sector, IoT is called IIoT (Industrial Internet of Things) or Smart Manufacturing or Industry 4.0. Here, IIoT means the use of IoT technologies to increase the competition and efficiency of manufacturing and supply chain operations. The interesting part is creating an environment that would better align with smart manufacturing decisions. In a survey conducted by Industry Week, it is stated that 40% of manufacturers think that this is the right time to invest.

Healthcare

The implementation of the Internet of Things in the healthcare industry is impressive and so immense. Industries will be able to extract data from medical devices to improve services for their patients. It will take the healthcare solution to all new levels of efficiency, precision, monitoring and patient safety, real-time visibility and intelligence will reach the data generated by sensors, portable devices and other IoT devices. Now, actionable insights into data are key to taking the organization to another level. Almost 60% of healthcare organizations have adopted IoT devices in their healthcare solutions and have significantly saved their savings graph. Internet of things (also known as IoT) IoT is a massive network of devices connected to the Internet and / or to each other. This can include anything with an on / off switch - computers, printers, cell phones, headsets, coffeemakers, televisions, washers, lamps, security systems, automobiles - the options and opportunities are virtually endless.

5.3 IoT System Management

By impacting everything we do, do, buy and sell, the IoT has the potential to mutually benefit both businesses and consumers in a substantial way: businesses will gain visibility and awareness, and consumers will gain empowerment. Let's examine some of the main ways that marketers can use the IoT to their advantage:

1) Increased Personalization and Insight into Consumer Behavior

As the IoT expands further, indoor location technology will allow marketers to connect with and interact with consumers in "real time" during their in-store experiences, even more than they already do. For example, marketers can tell when a customer has been behind a product for a while, and can send messages to the consumer's smartphone, encouraging the sale or offering a promotion. This type of restricted personalization will only increase over time, providing more insight into consumer behavior. Sellers will get a better idea of what exactly the consumer is doing, as well as when and why they are buying or not buying.

2) Customer Intimacy

Along with greater personalization comes greater intimacy between you and your clients. IoT technologies like social media and cloud computing provide consumers with the opportunity to provide instant brand feedback. In turn, responding quickly to consumers is an ideal way to increase intimate connections with them.

3) Focused Targeting

With the IoT providing a better understanding of consumer behavior patterns, marketers will be able to focus their efforts on a more specific goal, personalizing the experience for each user. So let's take a closer look at this. Marketing specialists can use IoT data to study consumers and target a specific audience; They can communicate directly with their target audience while improving customer service. This enhanced customer service can only help increase sellers' conversion possibilities.

4) Advanced Social Media Efforts

Because the IoT is optimized for use with social media, marketers will be better able to predict the development of social communities and target their efforts towards these communities. This will allow marketers to reach more leads than ever. IOT is a combination of different technologies integrated together. As in IOT, the signal (data) flows from the sensor to the cloud.

5.4 IoT Project

Occasionally known as the Industrial IoT, the business benefits of the IoT depend on the particular implementation, but the key is that companies should have access to more data about their own products and their own internal systems, and a greater ability to make changes. as a result. IoT thrives on improving transportation and production logistics. Helps track inventory, supply, demand, and even the consumer experience. It is also extremely useful in the development of Smart Cities, where it allows the monitoring of everything from public transport to waste management. It shows immense potential in the healthcare field, as drug monitoring devices that could ensure that patients are taking the correct doses of medications as well as remotely measure the patient's vital signs. Another domain where IoT is gaining popularity in agriculture and meteorology, where it allows you to track and collect information in remote locations.

5.4.1 The challenges of Developing an IoT Project

Most of the challenges of developing an IoT project come from a lack of knowledge and preparation. Some companies choose to develop it on their own to cut costs and end up failing, which in turn increases the overall cost of development. Management often underestimates the complexity of the project and neglects the research, testing, and time required to succeed. Like any project, an IoT requires time and money to develop. So be prepared to invest both if you want to do it right.

5.4.2 Design Steps for Developing an IoT Project

There are some differences between IoT solutions. That distinction comes from the fact that the companies that develop them knew why and how their products can improve their process and / or the life of the consumer. It may sound trivial, but many have failed simply because their idea was cool and smart, though not particularly useful. So, when you discover your project's use case, you can start to develop an implementation strategy, set realistic goals, and in a nutshell do your best.

Step 1: Find the use case
Most common use cases for IoT are:
- Preventative maintenance
- Remote monitoring
- Asset tracking
- Environmental monitoring

- Data gathering

When you establish the use case and audience for your project, you should start researching the necessary software and hardware.

Step 2: Research software and hardware.
Take into account the location of the devices, the frequency with which they can perform maintenance, their geographical coverage, connectivity, etc. Then make sure you know what features you want it to have and how it should work.

Step 3: Assemble a Team
For your IoT project you will need
- Mechanical Engineers
- Electrical Engineers
- Software Engineers
- Q/A Testing Experts
- UX/UI Designers.

Step 4: Build a Prototype
Only after you've assembled a team or engineers, designers, and testers can you start building a prototype. Developing it allows you to discover the minimum parameters for your IoT project before full implementation.

Step 5: Quality Assurance
A prototype must be thoroughly tested (both its software and hardware). You build something, test it and then you can improve it. It's a crucial step to any project.

Step 6 Build and deploy your IoT project.
After you've improved and tested the hell out of your device you are ready to build and deploy your first IoT project.

Step 7: Practice makes Perfect
IoT projects need constant improvement and maintenance to justify their development. Plan ahead and reserve time and budget for this specific step because overlooking it may cost you the entire project!

Chapter-6: Multiple Protocols in IoT Domain

6.1 Architecture of IoT

The most basic architecture is a three layer architecture. It has three layers, namely perception, network, and application layers.

(i) The perception layer is the physical layer, which has sensors to detect and collect information about the environment. Detect some physical parameters or identify other smart objects in the environment.

(ii) The network layer is responsible for connecting to other smart things, network devices, and servers. Its features are also used to transmit and process sensor data.

(iii) The application layer is responsible for delivering application specific services to the user. It defines various applications in which the Internet of Things can be implemented, for example, smart homes, smart cities and smart health.

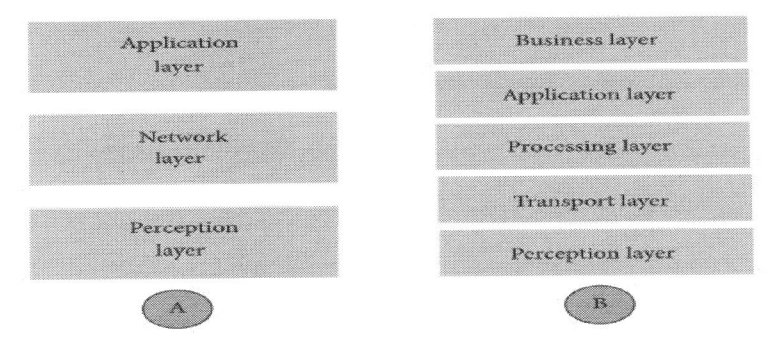

Figure 6.1: Architecture of IoT

One is the five-layer architecture, which also includes the processing and business layers. The five layers are the layers of perception, transportation, processing, application, and business. The role of the perception and application layers is the same as the three-layer architecture. The function of the remaining 3 layers is as follows:

(i) The transport layer transfers the sensor data from the perception layer to the processing layer and vice versa through networks such as wireless, 3G, LAN, Bluetooth, RFID and NFC.

(ii) The processing layer is also known as the middleware layer. Stores, analyzes and processes large amounts of data that comes from the transport layer. You can manage and provide a diverse set of services to the lower layers. It employs many technologies such as databases, cloud computing, and big data processing modules.

(iii) The business layer manages the entire IoT system, including applications, business and profit models, and user privacy.

The Internet of Things (IoT) enables communication between devices, commonly known as machine-to-machine (M2M) communication with the help of the Internet. Machine to Machine (M2M) is a broad tag that can be used to describe any technology that allows networked devices to exchange information and perform actions without manual human assistance. IoT works by connecting sensors and machines to collect the data and use it with cloud servers to interpret and transmit the data. IoT can be viewed in various categories, from handheld devices whose sensors will tell you your heart rate or how many miles you've walked or your location, to home care where a sensor in your air conditioner will tell you that you forgot to turn it off while leaving home and gives you access to turn it off from a remote location via your mobile phone, or the sensors in your car tell you that the tire pressure is low. IoT can be very beneficial in the healthcare area where a sensor in a human body will gather all your credentials like BP, Hb, etc. and will inform you if you need to go to the doctor. On a larger scale, it is used to determine the number of parking spaces, the quality of air and water, and the whereabouts of traffic.

6.2 IoT Architecture

IoT is a network that spans several elements and components: combined together, these components provide the services, commonly known as IoT services. There are several different parts of this network. There are sensors, actuators, gateways, transmission channels for short and long distance wireless / wired communications, back-end infrastructures, storage and cloud computing elements, mobility applications, and the like. For each of these items, there are several different options available on the market. The IoT architecture is basically an approach of how these various elements should be designed and integrated with each other, in order to offer a robust service delivery network that can meet future needs. Something similar to the architecture of our buildings, factories, roads, bridges, urban and commercial centers. Architecture is basically the backbone and must be carefully crafted with the evolving needs of functionality, scalability, availability, maintainability and similar criteria in mind.

The Internet of Things (IoT) is increasing at a rapid rate. Statista predicts that around 50B of IoT devices will be in use by 2030 worldwide. However, despite this massive presence of IoT and the business opportunities offered by this technology, a large number of people cannot understand this term. Furthermore, they are not familiar with the essential components necessary to build a reliable architecture for applications and implementations. Technically, IoT is a network of identifiable devices or objects that can communicate with each other through some type of connectivity, without human intervention. Let's analyze what elements are really necessary to create an interconnected infrastructure and empower physical devices with digital intelligence.

6.2.1 The Fundamental Layers of IoT Architecture

Since IoT systems are diverse in terms of industry-specific use cases, there is no viable architecture that can fully address all possible IoT applications. The most widely used IoT architecture comprises three layers:

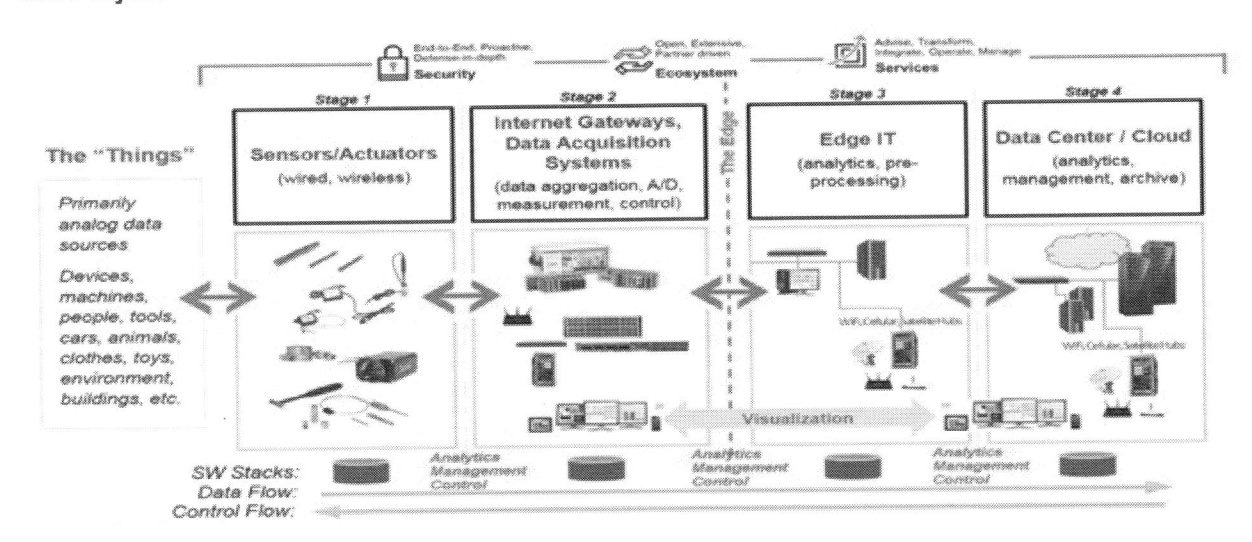

Figure 6.2: Four Stage IoT Solutions Architecture

The Perception Layer – The goal of this layer is to interact with the physical world through IoT devices integrated with sensors and actuators.

The Network Layer – This layer is intended to perform fundamental analysis of the data obtained by sensors using IoT gateways and communicate that data to a server for further processing through communication protocols such as MQTT.

The Application Layer – This layer provides an interface between the previous two layers related to hardware and business applications to provide application-specific services to users.

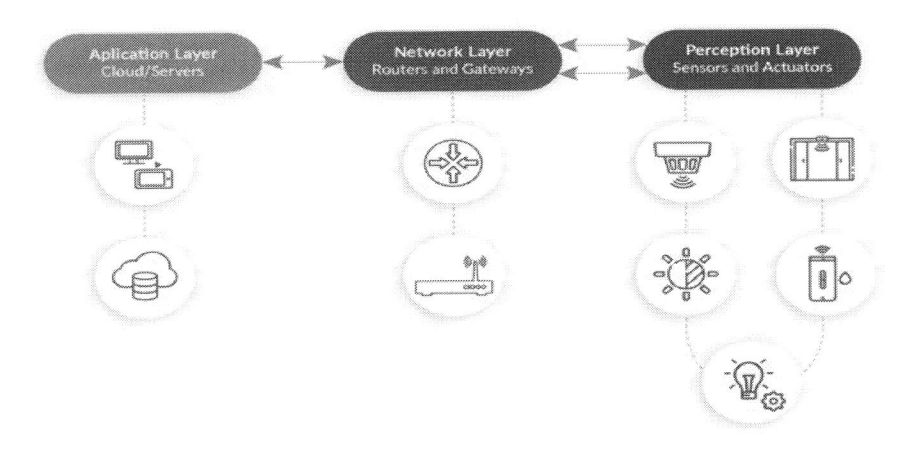

Figure 6.3: The Fundamental Layers of IoT Architecture

6.2.2 Key Components of IoT

IoT is much more than a set of elements such as actuators and sensors. Within the notion of IoT, devices and elements are connected in a single network or environment, with analytical and management systems, to facilitate automated device management, as well as data-based decision making. IOT offers you many services, such as data ingestion, dashboarding, platform integration, security services, data transformation, etc. Basically there are three layers of IOT architecture:

- IoT Device Layer: Sensors collect data from the environment and then measurement and convert it to useful data.

- IoT breakaway layer: The data coming from the first layer must be prepared before it enters the final processing stage.

- IoT Platform Layer: is to ensure the transition of data between field gateways and central Internet of Things servers.

IoT is a multilayer technology that is based on various elements and aspects that must be properly structured and connected to interact with each other and have an external impact. Such complexity describes the need and value of an IoT architecture model. The Internet of Things has already passed that development mark where its achievements can be confidently integrated in various fields and fields without taking into account its challenges and importance. From manufacturing and mining to Smart Home and City, the space and processes in which we live and participate have already begun to change.

Despite the proven potential for global distribution and integration of the IoT hardware architecture, it has a learning and integration curve for its adopters that slow down digital transformation across fields and its contribution in particular. It is about creating an ecosystem within an established environment must be properly adjusted to. It also involves a robust architecture to ensure its independence and stability, as well as regular monitoring and maintenance to ensure uninterrupted operation and perfect fulfillment of the initial purpose.

6.3 Basics of IoT Architecture

Despite all the variety of applications, there are not a lot of IoT architecture types that share some structural elements. It starts from the most-known part of the systems - embedded or separated devices. More exactly saying, their sensors and detectors that is responsible for the gathering of the required information about the environment. That data then should be passed through the next part of the structure. For the collected data acquisition should exits gateways or similar systems that should be able to handle the dedicated data flow - perform a digital conversion, filtering and preprocessing to provide the material for the following architecture block. The essential stage of the operation of any IoT ecosystem is data processing and analysis before it can be presented to users in a suitable form. This part is also where the system enters other advanced technologies like AI, machine learning and data science which approaches raising the level of scrupulosity and comprehension of the mentioned processes. The next block is the basis of the whole system - the infrastructure for the secure storage, appropriate management and more complex analysis of the necessary type and amount of data. There are four possible options for the issue - on-premise, cloud or edge computing and hybrid.

6.3.1 IoT Architecture Layers

There are three main stages that data must go through to ensure that the purpose of the system is met: sensor - gateway - infrastructure.

1) Devices

Any technology that enables digital transformation is driven by data that must be captured accurately and in a timely manner. Devices or objects equipped with sensors allow the measurement or detection of various physical parameters and telemetry that is of value for future research. These possibilities allowed IoT to enter non-technical fields such as agriculture to help it, along with others, enter the digital age and open a new era in development. For example, in this area, it allows centralizing monitoring and accelerating the reaction to unwanted changes and minimizing the possibilities of their appearance. However, sensor systems are not complete without actuators that transform data into necessary actions without human supervision. It allows complete automation of the different routine physical processes, improving their precision and long-term regularity. Allows outsourcing of machine procedures in difficult, even life-threatening, conditions.

Each collection of sensors can have a certain topology that improves the bidirectional interaction between the elements. In some cases, the appropriate level of connectivity between devices is even more vital than between acquisition systems and gateways. The important thing is that it must be supported in real time without revealing the need for management solutions. One of the challenges of the layer is balancing the needs of the devices (resources required for proper operation, including power, bandwidth, etc.) and the purpose of the entire system. That is why the robust system must be combined with the appropriate options for IoT architecture standards and communication protocols to ensure efficiency during the available periods.

2) Gateway and Data Acquisition

The necessary connection point between devices and data centers. Perhaps it could not be considered as a full layer rather than an intermediate extender. However, the stage is important for data collection and transmission. It allows to avoid the overload of the system, to filter unnecessarily and to pass each type of data to the required databases. Other layer roles are to convert data into an appropriate form, transfer and process it to other components of the architecture, and secure it to avoid read loss, corruption, and even attack. The vulnerability of IoT systems is one of the biggest challenges of scope today. That is why the data must be encrypted as soon as it takes a digital form.

3) Infrastructure

The type of data storage depends on the purpose of the system and commercial possibilities, as each one requires different maintenance efforts and features related to speed, storage, structure, access and security. On-site centers valued for security, in the cloud for their low cost and easy approach, while being innovative: flexibility and quick response. According to the market growth trend ($ 3.24 billion by 2025), the last option gains appreciation these days. It reduces the location of data sources compared to the cloud, which accelerates data flows and reduces the network, initially increasing its security and reducing the consumption of energy, bandwidth and other resources necessary for proper operation. However, cloud systems do not quickly lose their consumers, as they allow more massive volumes of data to be handled and their analysis more comprehensive than perimeter computing. Its value is marked by increasing adoption by the stream titled Industry 4.0 which establishes a high level of responsibility in IoT network devices and architecture. Use cases show that IoT integration brings manufacturing and production closer to improving profitability and accuracy. The same trend is shown in healthcare, supply chain, etc. On the other hand, the technology is quite easy to use, allowing supply companies the required intelligence through ERP and similar systems and its visualization through dashboards and reports.

4) Platforms and Applications

Although it can be considered as one of the components of the IoT architecture, it is a solution for managing the entire system that varies according to its topology, layer specifications, users, etc. It can come in the form of advanced software that includes extensive analysis. and predictive forecasting or limited to remote control of system operations. The IT industry makes its contribution to scope so that solutions can be accessed across all available platforms (web, mobile, desktop, or targeted) without significant effort. Considering that any IoT use case could be taken as unique, custom systems development has a large market share along with ready solutions from world leaders like IBM, SAP and Xiaomi. Every IoT architecture development like any task on the company's long list of digital transformation begins with research and analysis of existing conditions and possible options to find a secure and scalable formula for current and future needs. Based on efficient interaction, IoT is all about the right compatibility and connectivity to be able to fulfill its purpose

6.4 Common Issues and Threats of IoT Architecture Layers

Because IoT promises great opportunities, many organizations are looking to include IoT products in their business operations. Actually, it seems complicated to implement multiple devices and conditions necessary for it to work. That means that the problem of organizing a reliable IoT architecture naturally comes into the picture. To deal with the different factors that affect the IoT architecture, it is simple and more efficient to find a reliable IoT solution provider. There is no single, comprehensive agreement with IoT Architecture agreed by everyone and researchers. The researchers have proposed a different architecture. Due to the improvement in IoT and the challenges in IoT regarding security and privacy, IoT 4 layer architecture has also been proposed. Because of all these issues, use this guide to understand what happens during the IoT architecture. Before going any further, it is important to understand what this concept really means. Actually, IoT Architecture is the system of countless elements like sensors, actuators, protocols, cloud services and layers. In addition to that, IoT Architecture layers are differentiated to assess system consistency. And you need the same plan that any technology needs, including a plan for how it will merge with an organization's existing infrastructure and systems. Basically there are three layers of IoT Architecture including:

- IoT device layer (mainly client side)
- IoT Gateway Layer (server side operations)
- IoT platform layer (acts as a bridge to connect clients and operators).

Addressing the requirements of all layers is critical at all stages of the IoT architecture. On top of that, the core features of IoT Architecture involve functionality, availability, and ease of maintenance. If we are not addressing these conditions, the result of IoT Architecture will be a failure. The following are the main layers of IoT that provide the solution for the IoT architecture

6.4.1 Application Layer

The application layer defines all the applications in which IoT has been implemented. It is the interface between the final IoT devices and the network. IoT applications like smart homes, smart health, smart cities, etc. You have the authority to provide services to applications. The services may be different for each application due to the services based on the information collected by the sensors. It is applied through a dedicated application on the end of the device. As in the case of a computer, the browser applies the application layer. It is the browser that runs application layer protocols like HTTP, HTTPS, SMTP and FTP. There are many concerns at the application layer, of which security is the key issue.

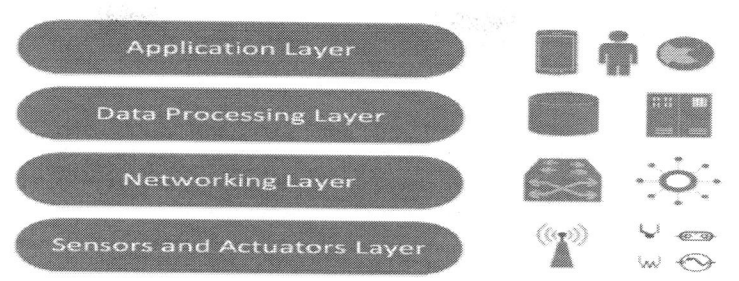

Figure 6.4: IoT Architecture Layers

Common Issues and Threats of Application Layers are:

1. Cross-site Scripting: It is a type of computer security diseases that are generally found in web applications. It allows attackers to inject client-side scripts, such as JavaScript, into web pages viewed by other users. By doing so, an attacker can completely change the content of the application according to their needs and use the original information illegally. The first method you can use to prevent scripts from appearing between sites in your applications is to escape user input. Escaping the entry means taking the data from an application that has received and guaranteed its security before providing it to the end user. Entry validation is another process to ensure that an application provides the correct data and protects harmful data from damaging the site, database, and users.

2. Malicious Code Attack: This is a particular code in any part of the software or script system that is considered to cause undesired effects, security threats or damage to the system. It is this threat that cannot be blocked or controlled by the use of antivirus software. Static Code Analysis (SCA) is an effective method of protecting malicious code from causing harm to computers. SCA is a method of debugging a computer program that is performed by analyzing the code without running the program. The process provides an understanding of the structure of the code and can help ensure that the code must match industry standards. Today's top scanners can easily detect malicious code such as data leak, time bombs, anti-debugging techniques, back door threats etc.

6.4.2 Data Processing Layer

In the three-layer architecture, the data was sent directly to the network layer. Due to sending data directly, the chances of damage are increased. In the four-layer architecture, data is sent to this layer, which is derived from a perception layer. The Data Processing Layer has two responsibilities: it confirms that authentic users send the data and they avoid threats. Authentication is the most widely used method to verify users and data. It is applied through the use of pre-shared keys and passwords for the user in question. The second responsibility of the layer is to send information to the network layer. The medium through which data is transferred from the Data Processing Layer to the network layer can be wireless and cable based. Common problems and threats in the data processing layer are:

1. **DoS Attack**: An attacker sends a large amount of data to overload network traffic. Therefore, the large consumption of system resources depletes the IoT and renders the user unable to access the system. Implementing an antivirus program and a firewall will restrict bandwidth use to only authenticated users. Server configuration is another method that can help reduce the likelihood of being attacked.

2. **Malicious Insider Attack**: It comes from inside an IoT environment to access private information. It is performed by a user authorized to access the information of another user. Practices such as data encryption, implementing good password management practices, and installing antivirus will help keep your data safe from such threats.

6.4.3 Network Layer

This layer is also known as a transmission layer. It acts as a bridge that transports and transmits data collected from physical objects through sensors. The medium can be wireless or cable based. It also connects network devices and networks to each other. Therefore, it is extremely sensitive to attacks by attackers. You have significant security concerns regarding the integrity and authentication of data transmitted to the network. Common network layer problems and threats are:

1. **Main-in-The-Middle Attack:** The MiTM attack is an attack in which the attacker privately intercepts and modifies the communication between the sender and receiver that they assume are communicating directly with each other. It leads to a serious online security threat because it gives the attacker the way to capture and control data in real time. Secure / multipurpose Internet mail extensions encrypt emails ensuring that only intended users can read and will avoid data from MITM attacks.

2. **Storage Attack:** Crucial user information is saved on storage devices or in the cloud. Both the storage device and the cloud can be attacked by the attacker, and user information can be modified to obtain incorrect details. By regularly backing up files, running antivirus software, and using a system with strong passwords to keep data access restricted, we can protect the data from the attacker.

3. **Exploit Attack:** An exploit is any unethical or illegal attack in the form of software, data blocks, or a script. Take advantage of security deficiencies in an application, system, or hardware. It usually occurs in order to gain control of the system and steal information stored on the network. By installing all software patches, security releases, and all updates to your software, there are few preventative measures against attacks.

6.4.4 Perception layer/Sensor layer

The sensor layer has a responsibility to recognize things and collect data from them. There are many types of sensors attached to objects to collect information, such as RFID, sensors, and 2D barcodes. Sensors are selected based on application requirement. The data collected by these sensors can be about location, changes in the air, the environment, etc. However, they are the primary target of attackers who want to use them to replace the sensor with their own. Therefore, most threats are related to sensors are

1. Listeners: This is a real-time unauthorized attack in which an attacker intercepts personal communications, such as phone calls, fax transmissions, and text messages. It tries to take crucial information that is transferred over a network. Preventive measures such as access control, continuous monitoring / observation of all devices, and thorough inspection by a qualified specialist in technical countermeasures of all components are necessary.

2. Playback Attack: Also known as a Playback Attack. It is an attack in which an attacker intrudes on the conversation between the sender and the receiver and extracts authentic information from the sender. The added risk of replay attacks is that a hacker doesn't even need enhanced skills to decrypt a message after capturing it from the network. This attack can be avoided by using time stamps on all messages. This protects hackers from forwarding messages sent longer than a specified period of time. Another preventive measure to avoid becoming a victim is to set a password for each transaction that is only used once and is discarded.

3. Time Attack: Generally used on devices that have weak computing capabilities. It allows an attacker to find vulnerabilities and remove secrets stored in the security of a system by observing how long it takes for the system to respond to various queries, entries, or other algorithms. To avoid this attack, we can simply do a constant time comparison using the timer functions. You need to test to make sure compiler optimizations don't put time issues back in your code.

Chapter-7: Common Security Measures used for Designing an IoT Applications

7.1 Introduction

We need to understand the entities included in this definition. Let's start with "Things" for an integrated developer; it is easy to identify them with integrated devices. This definition includes small microcontroller-based devices, as well as more complex devices running a full-featured operating system. Here are the six main attributes that make "things" part of the Internet of Things, or IoT:

Sensors - IoT devices and systems include sensors that track and measure activity around the world. An example is Smartthings opening and closing sensors that detect whether a drawer, window or door in your home is open or closed..

Connectivity: Internet connectivity is contained in the item itself or in a connected hub, smartphone, or base station. If it's the latter, then the base station will likely collect data from an array of sensor-laden objects and transmit data to the cloud and vice versa.

Processors: Like any computing device, IoT devices will contain some "hidden" computing power, if only to analyze incoming data and transmit it. All of these features apply to today's smartphones, of course, but many IoT devices will also need to be equipped with various special features to be truly useful. These will differentiate IoT devices, particularly remote ones, from today's smartphones.

Energy Efficiency: Many IoT devices can be difficult, expensive, or dangerous to access to charge or replace the battery. You can even think of the Mars Curiosity Rover as an example of such a device. Therefore, they may need to be able to operate for a year or more unsupervised using a conservative amount of energy or be able to wake up only periodically to transmit data.

Cost-effectiveness: Objects containing sensors may have to be widely distributed to be useful, as in the case of sensors in food products in supermarkets that would indicate whether an item has been damaged. These would have to be relatively cheap to buy and implement.

Quality and Reliability: Some IoT devices will need to operate in harsh outdoor environments and for long periods of time.

Security: IoT devices may need to transmit sensitive or regulated information, such as health-related data, so data security will be critical.

The "Internet of Things" may just be about connecting devices to the Internet. It would be like defining the Internet as PCs, tablets and smartphones connected to a TCP / IP based network; we would miss an important part of this. We already have many connected devices. Multiple studies assess that, since 2008, there are more devices than people connected to the Internet, but the IoT is still far from being a reality in our daily routine. The Internet has evolved in recent years, from the static "email and website" model at the start of the World Wide Web revolution to the huge and sometimes useful combination of information and services (and cat videos) we have today. . Similarly, connected devices will need to evolve, interact, and share and access information before they can truly live in the "Internet of Things" era.

The "Internet" side of IoT is about processing the large amount of data that devices can collect and extracting useful information that can improve the way we use many services and devices today. This may sound like finding a needle in a haystack, but the good news is that with cloud computing we have the processing power we need to filter out many haystacks per second. Data management is a crucial aspect of the Internet of Things. By considering a world of interconnected objects and constantly exchanging all kinds of information, the volume of data generated and the processes involved in handling that data become critical. There cannot be an "IoT device" (it will be just a "thing"), but we can design a device that is part of an IoT solution. The device will no longer be "the product", but will be part of a larger project that involves services that add value. We also have processing power on devices, and IoT does not mean that all processing must be done in the cloud, but on the other hand, having a complete view of the set of devices can be very useful to be more informed and more efficient decisions. The Internet of Things means taking all the things in the world and connecting them to the internet.

7.2 IoT Matters

When something is connected to the Internet, that means you can send information or receive information, or both. This ability to send and / or receive information makes smart and smart things good. Let's use smartphones (smartphones) again as an example. Right now you can listen to almost any song in the world, but it's not because your phone actually has all the songs in the world stored. This is because all the songs in the world are stored elsewhere, but your phone can send information (asking for that song) and then receive information (transmit that song on your phone). To be smart, a thing doesn't need to have super storage or a super computer inside. All you have to do is connect to a super storage or connected supercomputer is amazing. In the Internet of Things, all things that connect to the Internet can be classified into three categories:

1. Things that collect information and then send it.

2. Things that receive information and then act on it.

3. Things that do both.

And all three have enormous benefits that feed each other.

7.2.1 Collecting and Sending Information

This means sensors. The sensors can be temperature sensors, motion sensors, humidity sensors, air quality sensors, light sensors, whatever. These sensors, together with a connection, allow us to automatically collect information from the environment, which, in turn, allows us to make smarter decisions.

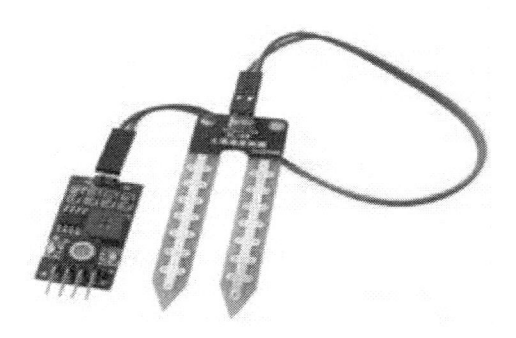

Figure 7.1: Soil moisture sensor

On the farm, automatically obtaining information on soil moisture can tell farmers exactly when to irrigate their crops. Instead of watering too much (which can be costly overuse of irrigation systems and environmental waste) or watering too little (which can be costly crop loss), the farmer can ensure that crops get exactly the correct amount of water. More money for farmers and more food for the world! Just as our sight, hearing, smell, touch and taste allow us humans to make sense of the world, sensors allow machines to make sense of the world.

7.2.2 Receiving and Acting on Information

We are all very familiar with the machines that get information and then act. Your printer receives a document and prints it. Your car receives a signal from your car keys and the doors open. The examples are endless. Whether it's as simple as sending the "activate" command or as complex as sending a 3D model to a 3D printer, we know we can tell machines what to do from afar. And that? The true power of the Internet of Things comes when things can do the above. Things that collect information and send it, but also receive information and act on it.

7.3.3 Doing Both

Let us quickly return to the example of agriculture. Sensors can collect information on soil moisture to tell the farmer how much to water the crops, but it doesn't really need the farmer. Instead, the irrigation system can automatically turn on as needed, depending on the amount of moisture in the soil. IoT gives businesses and people better insight and control over 99 percent of objects and environments that remain outside of the Internet. And by doing so, IoT enables companies and individuals to be more connected to the world around them and to do more meaningful and higher-level work.

7.3 Common IoT Security Measures

Embedding security in the design phase: IoT developers must embrace security at the start of any consumer, business, or industry-based device development.

PKI and Digital Certificates: Public Key Infrastructure (PKI) and 509 Digital Certificates perform essential functions within the development of secure IoT devices, providing trust and management.

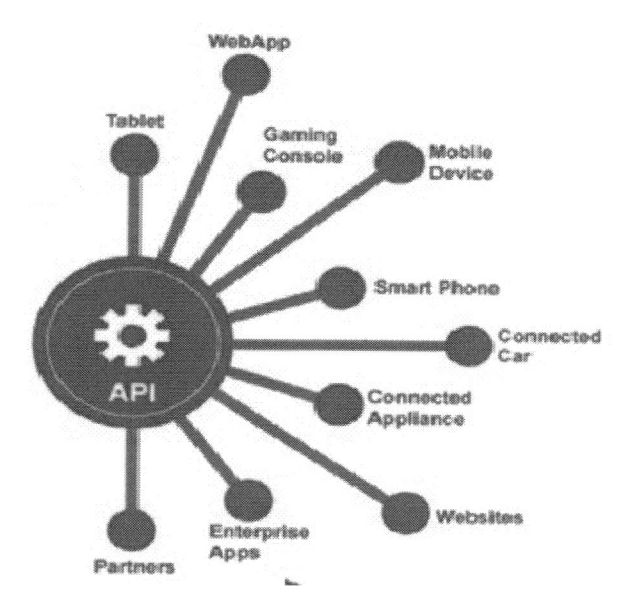

Figure 7.2: API Security

API security: Application performance indicator (API) security is important to protect the integrity of the information that is sent from IoT devices to back-end systems and to ensure only approved devices.

Hardware security - This is particularly vital once devices are used in harsh environments or where they are not to be physically monitored.

Network Security: Associate-grade IoT network protection includes ensuring port security, disabling port forwarding and never-needed-range ports; victimization anti malware, firewall and intruder detection system / intruder bar system; interference from unauthorized scientific directions; and warranty systems are patched and updated so far. Connecting vehicles, household appliances, microchips, medicinal equipment using integrated electronic devices, etc. to collect and market data of a different type is called the Internet of things. This technology allows the customer to control devices remotely through a system.

7.3.1 IoT Examples

Some of the IoT examples in Real Life Implementation are:

1. In wearable technology: wearable devices like Fitbit bands and Apple watches sync easily with mobile devices. These also help to display information, cell phone warnings on them.
2. Infrastructure and development: With the use of an application, it is easier to obtain constant information on outdoor lighting and, depending on this, the street lights turn on or off.
3. Health care: there are various applications to control the health conditions of patients. Based on baseline data, administrations monitor the medication dose multiple times on a multiple day.

7.3.2 Technology used in IoT

- RFID (Radio Frequency Code) and EPC (Electronic Product Code) tags
- Z-Wave is a low power RF communication technology used for home automation, lamp control, etc.
- NFC (Near Field Communication) is used to allow two-way interactions between electronic devices, which is generally for smartphones and is mainly used for contactless payment transactions.
- WiFi is the most widely used option for IoT. This helps to transfer files, data and messages smoothly when on the LAN.
- Bluetooth is used where short-range communications are enough to get away with it. This is mainly used in portable technologies.

7.3.3 Testing IoT

When testing any IoT system, we must go through several testing approaches:

1) Usability
2) IoT security
3) connectivity
4) performance
5) Compatibility tests
6) Pilot test
7) Regulatory tests
8) Update test

7.4 IoT Testing Challenges

- **Hardware-Software Mesh**: IoT is an architecture that is tightly knit between different hardware and software components. It is not only software applications that create the framework, but also hardware, sensors, correspondence gates, etc. they also play an important role. Thus, it becomes a tedious activity compared to testing a conventional system (software / hardware component only).

- **Device Interaction Module**: As it is architecture between several sets of hardware and software, it is mandatory that they talk to each other in real time. When both integrate with each other, things like security, reverse compatibility, upgrade issues become a challenge for the test group or team.

- **Real-time Data Testing**: Being in a test group or team, obtaining administrative checkpoints, or deploying the system on the pilot is very intense. The step turns out to be much more difficult if the system is related to healthcare more or less. In this way, that remains a great challenge for the test group or team.

- **UI**: The IoT is spread across devices that belong to all platforms (Android, Linux, iOS, Windows). Now testing that on devices should be possible, however testing it on all possible devices is relatively impossible. We can't ignore the likelihood of the UI being accessed from a device that we don't have or simulate. That is a difficult challenge to overcome.

- **Network Availability**: Network connection plays a crucial role, as IoT is the information that is constantly delivered at faster speeds. The IoT architecture must be tested over a wide range of network speeds / connectivity. To test this, virtual network simulators are generally used to change network load, stability, connectivity, etc. But real-time data / network is always another situation and the test group or team doesn't know where the bottleneck would be created in the long run.

There are several tools that can be used to overcome challenges during IoT testing. For the creative world around us, IoT is a developing market and has great opportunities. The time is not far off when IoT becomes critical for evaluators to survive in the development world. The IoT device, the smart device application and the correspondence module play an important role in contemplating and evaluating the performance and conduct of the different IoT services.

7.5 IoT Product Developments
In 2011, Marc Andreessen stated that "software is eating the world," placing a heavy emphasis on companies running on smart software and greatly disrupting many industries. Today, as we become

gadget-obsessed users whose lives are surrounded by those devices that run on a smart range of software, that statement could no longer be true. The Internet of Things industry is flourishing, with a projection of 20 billion connected devices in the world by 2020, according to a Gartner report. In this ultramodern world, we see a plethora of IoT devices, rolled up every day to make a significant change in our lives, from smart plugs to smart toothbrushes, irrigation controllers to Wi-Fi lighting, smart thermostats, and power cameras. security connected. Unquestionably, IoT technology is permeating every aspect of our lives and we cannot allow separation from the growing connected world.

Considering IoT penetration at blazingly fast speeds, Bellwether companies are striving to harness the right mix of talent to work on IoT products that unequivocally permeate the lives of users. Today's software developers should have to reach out - they must be a quick troubleshooter who can chart the roadmap, recognize that the ecosystem is constantly changing, and realize that their responsibility has to be kept up to date. IoT is itself a broad term that includes security, networking, systems engineering, cloud programming, and obviously hardware device programming. An IoT developer must be multilingual so that he can be flexible and play a multiple role on the team. Generally speaking, there are four main stages of IoT product development:

Physical Hardware: This stage requires engineering skills that are far beyond the scope of the software developer. Most IoT products use pre-mounted boards and sensors built into these boards.

Device Programming: In this case, programming skills play the primary role in reading data from connected sensors on the IoT device and sending it to the cloud server.

Programming the server that stores product data: Well, this stage involves the use of server-side languages like PHP, ASP.NEY, Javascript, Node.js and database queries mainly in MySQL or other alternatives.

Show data to product user: This stage includes creating a web console or web page to display the stored data to the user, which again requires skills like PHP, JavaScript, HTML, CSS, MySQL or other frameworks.

7.6 Steps for IoT Product Developments

Deep understanding of sensors - A talent pool who would like to work in the IoT domain should be well versed in sensors and wireless communication. IoT developers are highly recommended to have a background in computing or electrical engineering. If you really want to rock the show, start with online courses on sensors and project development. A series of low-priced sensors and manufacturer's plates / kits are available to start the project on your own. There is no doubt that IoT will take you into the world of mechanical and civil engineering as the sensor receives physical data over a wide range.

User Interface: While developing an IoT-based product or device, it is imperative to follow high-quality standards for user experiences. In this fierce competition, an ambiguity in the intuition or usability of the IoT product can generate mistrust in the user. As part of a leading IoT company, we must ensure that the product meets the desired quality standards and that it delights technology-savvy customers. Product quality and reliability must be maintained and usability studies are a must for all IoT developers for this.

Learn JavaScript / Python: It is imperative for an IoT developer that he / she is proficient in JavaScript or Python language. Using the skills of such web programming languages is good for both the data processing backend and the code running on the device itself. Being an event-driven language, Javascript is ideal for reacting to new data that is continually coming from devices and triggering actions on the device at the same time. Using common language including Python and JavaScript with some optimistic Windows IoT devices using C #, .net languages would make sense for IoT projects.

Kick-Start with Raspberry Pi: Let me shed some light on Raspberry Pi strengths. It is a small computer in itself and is quite often used in creating proofs of concept for IoT based projects. This small size computer also provides a great way to learn how to link / solder simple circuits, and thus software circuits. For example, using a device called Tessel 2 or Particle Photon, or even simple Raspberry Pi can make developers ready to tackle the IoT space and learn how hardware works and learn new skills.

In fact, it's a great way to explore, develop, and refine ideas and then bring them to life. The IoT world is still in its infancy, so there are some well-defined paths to start with. However, it offers untapped opportunities and enormous benefits for those who are eager to explore new things going beyond their limits. Also, one must remember that technology continues to change regardless of whether it's sensors, single-board computers, or other embedded platforms. IoT developers should have to stay flexible to adapt the changes and move on in this generous world of IoT.

Chapter-8: Python Logical Design of an IoT system

8.1 Python in Internet of Things (IoT)

The rapidly changing automotive industry has enabled the IoT to revolutionize the automotive industry. The Internet of Things (IoT) makes driving safe and efficient. It has unleashed a number of benefits in agriculture, from improving productivity to the risks of crop failure. The IoT's ability to diagnose a problem and prevent system failure is helping to prevent the failure scenario. The use of IoT devices has increased year after year. Over 8 billion IoT devices were registered from 2016 to 2018. According to the analysis by the IoT expert, by the end of 2020, the IoT device count will reach more than 30 billion. And the IoT market value will reach $ 7 billion. As the Internet of Things (IoT) continually evolves, it can be difficult to analyze which tools are most useful in IoT development. Many programming languages are used to develop IoT devices. But which programming languages are more efficient in IoT development. The Python language is one of the most popular programming languages for IoT. The coding flexibility and dynamic nature of python help developers create smart IoT devices.

8.2 Python

Python is a very popular high-level programming language that focuses on code readability. It is a dynamic and interpreted programming language. Python supports multiple programming paradigms. In general, Python has fewer steps compared to Java and C. The Python language is also called a general-purpose programming language. Python can be used for software development, math, web development, and system scripts. In Python web development, developers use server-side python to develop web applications. Python can handle large and complex data easily. It can work on different Windows, Mac, Linux and Raspberry Pi platforms. Python is an efficient and fast programming language because it runs in the interpreter. Python can be treated as procedural, functional and object oriented, etc. With the scripting language, you can develop desktop applications and web applications. It was also translated into the binary language like Java.

8.2.1 Python in IoT Development

A database is obvious when it comes to most IoT applications. All IoT devices send data to the internet. So there should be a required database that can store the generated data. MySQL is the relational database for most developers. In this sense, mysqldb is a very convenient little tool that avoids the need to run shell commands inside a Python script to read and write to a database. And Python enters the image. You can also use other programming languages along with Python like:

- Assembly
- C
- C++
- Java
- Javascript
- PHP
- Python

Previously, developers used a java programming language in IoT development. Python is a developer's favorite language today. The reason behind the use of Python in IoT development is the specific feature that Python provides:

- Easy to code: with clear syntax, developers have an idea about code identification instead of {};
- Simple Syntax: Python has a simple syntax similar to the English language
- Interpreted language: Python runs on the interpreter system. The code can be executed as soon as it is written. Prototyping can be very fast.
- Integrable: Python allows integration with other languages. It is possible to put our code in other programming languages like C ++, etc.
- Extensible: Python is an extensible language. It allows developers to write programs with fewer lines than some other programming languages.
- Portable: Python code is portable, there is no need to change the code for different machines. You can run a code on many machines
- Free and open source: Python is an open source language. Its source code is freely available to the public, you can download, change and distribute it.
- Community Supports: Python has already gotten its great response in the market with the features mentioned above, so it provides many users grouped in the community to further support the advances.
- Easy to Learn: Learning and implementing Python is relatively simple and easy compared to other native languages like C ++ and Java.
- Easy to debug: The Python scripting language is one of the best for debugging C ++ and C. This source code runs line by line.
- Library support: Python supports large standard libraries. Installation of the libraries is easy, and it saves time.

8.3 Advantages of Python for IoT

- Python's popularity is a considerable asset: the language is backed by a large and useful community, which has led to the creation of a comprehensive set of prewritten libraries, making it easier to deploy and deploy working solutions.

- Python is portable, expandable, and embeddable : this makes Python system-independent and allows it to support many of the single-board computers currently available on the market, regardless of architecture or operating system.

- Python works great for managing and organizing complex data: for IoT systems that are particularly data heavy, this is especially useful.

- Python is easy to learn without forcing you to familiarize yourself with many formatting standards and compilation options: the most immediate consequence of this is faster results.

- Python code is easily readable and compact thanks to its clean syntax: this is useful on small devices with limited memory and computational power. Additionally, the syntax is partly responsible for Python's increasing popularity, further strengthening its community.

- Python's close relationship with scientific computing has allowed it to gain ground in IoT development: if a social scientist or biologist wants to create a program for their smart device in the lab, they will happily use their favorite language. In most cases, that will be Python, as it is the reference technology for scientific computing.

- Python is the language of choice for Raspberry Pi: It matters a lot since Raspberry Pi is one of the most popular microcontrollers on the market.
- Python offers tools that simplify the IoT development process, like webrepl: this gives you the option to use your browser to run Python code for IoT.

8.3.1 Raspberry Pi for Python in IoT
The Raspberry Pi:

- it is small (85mm × 56mm for Raspberry Pi 3);
- consumes very little energy;

It comes equipped with USB ports, an HDMI port, an Ethernet port, and Micro SD support. Most importantly, it has a Linux distribution on board, which means it also uses Python, making Raspberry Pi coding simple and straightforward. The Raspberry Pi is an incredibly versatile device that you can use to build anything:

- a media center,

- a retro game machine,
- a time lapse camera,
- a robot controller,
- an FM radio station,
- a web server,
- a motion capture security system,
- a Twitter bot,
- a mini desktop PC.

8.4 Project Based on Measurement of temperature and humidity Using IoT

8.4.1 Requirements of Things

- DHT11 IC: It is an IC temperature and humidity sensor. You can choose as many and any sensor you want is cheap and easy to start an IoT project.
- UNO Arduino Board.
- Basic knowledge of Python and Python IDE.
- API: APIs are nothing more than what you can understand as hyperlinks that you will use to send and obtain data. You can use Internet Of Things - Thing Speak in the cloud for API.
- Internet.

The DHT11 IC code is available on the Internet for Arduino so you can copy and paste it or if you want to create your own code you can use the datasheet to burn your code to the Arduino board. Now IoT is all about giving the sensor power so it can transmit data to another device or cloud. The reason I choose Python is because it is a very easy language if you don't know the coding, just read the basics of Python and it won't take long and to send data you can "request" the Python library to publish (send) and get data from the cloud. You can use the python "serial" library to get data from the arduino sensor to your laptop (where it has python IDE).

8.4.2 Example of Python Code

```
Import serial
Import requests
Arduino serial data = serial. Serial ('COM5', 9600) // COM5 is the port no (you can check it from device manager which arduino port it is running on, 9600 is the baud rate)
while (True):
if (arduinoSerialData.inWaiting ()> 0): // if data is available
print ("Available data")
myData = arduinoSerialData.readline () // reading sensor data
sensorData = float (myData) // making it float
```

h = request.post ("https://api.thingspeak.com/update?key=A0CHUDDxxxxxxxxx&field1=%f"% (Sensor data)) // post means we are sending data to the cloud. You will get your own key (url) once you create your account at thingspeak.com. You can verify the data you are sending to your account at thingspeak.

The same if you want to get your data from anywhere in the world, use

h = request.get ("https://api.thingspeak.com/update?key=A0CHUDDxxxxxxxxxxxx") // your API to get the

8.5 Project Based on ON/OFF LED or Interfacing Temperature Sensor or Speed Control of Motor Using IoT

Step 1: To get started with IoT, take a small definition like on / off LED or interface temperature sensor or motor speed control.

Step 2: To decide an ioT gateway to work (Arduino, ESP8266, Raspberry PI, banana PI, Orange PI), etc.

Step 3: To start making the app and use free public clouds like: Carriots, things talk.

Step 4: connect more devices to GPIO and start controlling them.

Three essential components are needed to build any IoT project. They are

1. Microcontroller: which acts as the brain of the system. Ex: Arduino and Raspberry Pi

2. Sensors: that collect data in real time and send it to the microcontroller. Ex: DHT 11, soil moisture sensor, water flow sensor

3. Communication module: that connects the system to the cloud for data storage and analysis. Ex: ESP8266 WiFi module

8.6 An IoT project to read a temperature sensor and send data to the cloud

1. Sensors: Sensors must be selected based on the application you want to create. E.g. Temperature sensor

2. Gateway: Gateway connects its sensors to the cloud. For example: Beaglebone, CC3200, Raspberry Pi, Linkit One, Intel Edison, Arduino Mega.

3. Cloud / Server: this is the end where you manage your data, perform analysis. You can build your own server / cloud from scratch, but if you want to focus on other aspects of your project,

you can use various cloud services provided by various providers. For example: IBM Bluemix, Microsoft Azure, AWS

Above all, you must decide which protocols to implement to send data from the gateway to the server / cloud. For example: MQTT, CoAp, Http.

Step 1: Decide on a project to give you hands-on experience. If you want to get started easily, you can choose something like: "read a temperature sensor (like DHT-11) and send data to the cloud"

Step 2: Choose a DIY platform like Raspberry PI Raspberry Pi or Arduino Arduino - Home, these platforms will help you read your sensor and send data to cloud

Step 3: Use a service like Internet Of Things - ThingSpeak and create a channel to start sending your data.

Chapter -9: Laboratory Companion for Designing IoT Applications

9.1 Internet of Things (IoT)

The Internet of Things (IoT) is ushering in a new era in science and technology that will forever change our personal and professional lives, our consumer habits, and the way we do business. With the world rapidly changing, these latest inventions and innovations will become the norm by 2020, and we estimate that more than 50 billion devices will be connected via the Internet. To create the first users, we have introduced a unique course on "Internet of Things", the next big thing in the IT industry. It is estimated that around $ 6 billion will be spent on Internet of Things (IoT) solutions in the next five years. There is no price to guess that the reason why you are expected to spend this amount on IoT is that IoT has already shown a lot of potential in a very short time and it has just started. There is not a single industry that is not affected by IoT for now. Some have already been heavily influenced by IoT, while others are just beginning to realize its importance. It is a collective term that defines a system of devices connected in a network. Being part of the network allows them to receive, save and exchange information among themselves. In a nutshell, the IoT architecture consists of several components:

1. **Sensors:** the purpose of these is to collect all the data from the environment for further processing. For example, sensors such as gyroscopes and proximity sensors act as "eyes" as they define the environment of the device. The gyroscope can help identify movement and rotation, while proximity sensors will indicate how close objects are surrounding the device.

2. **Data acquisition systems:** these are the devices that digitize the data collected by the sensors and also pre-process before the data can be sent to Stages 3 and 4.

3. **Edge IT:** This is the transit to the "IT realm". Digitized and aggregated data is collected and processed before sending it to the data center. This analysis can be performed by private IoT software services.

4. **Data Center:** This is where the collection date is obtained for full analysis and processing. Data from all types of sensors can be combined here for more detailed information. The data centers can be physical or cloud-based, created by the software development company IoT. The type of analysis remains the same, regardless of the platform provided.

IoT is a massive growth industry in today's times. Mastering a skill takes years of practice and perseverance. But a good start in the right direction is the most important aspect to be successful. Now, IoT has a 4-tier service architecture in which at each tier there are a large number of devices and services that communicate to create an integrated solution. Each of these layers consists of a technology / platform that you will have to learn and move forward to learn your skills. Here is a representation of the 4-stage architecture:

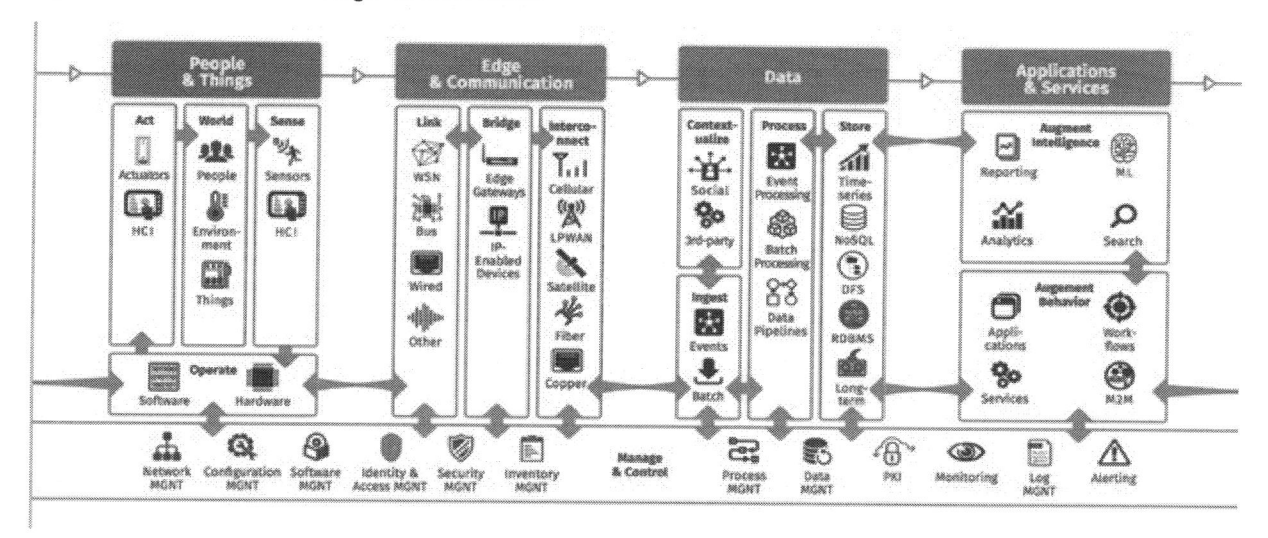

Figure 9.1: 4-Stage Architecture of IoT

Below is a concise list of all the topics you need to have basic ideas for jumping into the IoT world:

1. **Device Hardware:** You must have an idea of the architecture and operation of various microcontrollers (Arduino / RaspberryPi / Beagle Bone, etc.) along with various sensors such as temperature sensors, motion sensors, etc. You must be able to connect them all together on a complete electrical circuit.
2. **Device Software:** Being able to encrypt is the most important aspect of IoT. You should be able to write programs to be able to configure your controller and have it act accordingly. Start learning how APIs work within microcontrollers and how you can use built-in libraries for programming.
3. **Communication and cloud platform:** being able to send and receive data is the core of IoT. You must learn the basics of wired and wireless communication. Although Cloud itself is a very big topic, but it is an indispensable part of IoT. Maintain firm control over how Cloud technology works and your IoT integration.
4. **Cloud application and UX:** These two are not related to each other, but their objective is to serve a common purpose, improve and facilitate the user experience. Cloud applications are applications that run in the cloud and have faster and easier accessibility. The user experience or user experience also serves to enhance our ability to use the system to its fullest potential with ease.
5. **Security, Standards and Regulations:** These are the exploding technologies that have given the IoT boom. Each of these forms an IoT backbone or is itself an IoT product.

You should be thorough about how all of these technologies work and how IoT works in collaboration with each of them.

9.2 Advanced IoT

IoT extends computing and Internet connectivity from the most widely used devices, such as desktops and laptops, smartphones and tablets, to various variants of devices and everyday things. The data captured in these devices through sensors reveal interesting patterns with latent commercial values. Companies are increasingly interested in leveraging data-driven insights to create value. This amazing technology is creating tremendous opportunities for companies to reap greater benefits by improving resource efficiency and increasing productivity.

9.2.1 Internet of Things Learning Outcomes

- Expert knowledge of IoT technology and tools
- Strong understanding of IoT core concepts, background technologies, and subdomains
- Knowledge and skills of sensors, microcontrollers and communication interfaces to design and build IoT devices
- Knowledge and skills to design and build a client-server based network, and publish-subscribe to connect, collect data, monitor and manage assets
- Knowledge and ability to write scripts and applications from devices, gateways, and servers to aggregate and analyze sensor data.
- Knowledge and skills to select application layer protocols and web services architectures for seamless integration of various components of an IoT ecosystem
- Knowledge of standard development initiatives and reference architectures.
- Understanding the implementation of various types of analysis on machine data to define context, find flaws, ensure quality, and extract actionable information
- Understanding commercial and industrial cloud infrastructure, services, APIs and cloud platform architectures.
- Understanding of the predominant computing architectures: distributed, centralized, edge and fog.

IOT is actually a combination of different technologies, primarily sensors, communications, networking, cloud, analytics, and applications. In my opinion, these five are the key pillars (or say foundation) of IOT. So, below is a plan to learn IOT:

- **Sensors:** Sensors are used to collect large amounts of data. Different types of modern sensors allow us to collect data that was not possible before. This new data is giving new ideas and therefore better supervision and control of the underlying processes. RFID tags were the initial enablers to develop IOT systems.

- **Processor:** different processor architecture, interface and programming (specially integrated c)

- **Communication technologies:** sensors must communicate their data using different communication technologies such as Wi-Fi, Zigbee, NFC (Near Field Communications), etc. The use of any of these technologies depends on the application and the part of the system where it is being used.

- **Network topologies and protocols** - Huge no. sensors are implemented in different network topologies depending on the use case. Different types of protocols are required with different restrictions on power consumption, delay, physical area and distance, etc. MQTT is a popular IOT broker protocol that is based on the publish-subscribe architecture.

- **Cloud technologies:** all data must be stored somewhere to be analyzed. Now a cloud of days (nothing but digital space for rent) is the place for this. Most IOT systems have transmission data, so special types of databases are used for this. Analytical algorithms are used in this cloud data to infer or predict. Machine Learning and Artificial Intelligence algorithms are used in some advanced systems of the latest generation.

- **Applications (web or mobile)** - This is the user interface part of the entire IOT system. Applications are developed in IOT systems to display knowledge and prediction to the end user. In some IOT systems, applications are also used to monitor and control the system.

9.3 Arduino

Arduino is an open source electronic platform based on easy to use hardware and software. Arduino boards can read inputs (light on a sensor, a finger on a button, or a Twitter message) and turn it into an output: start a motor, turn on an LED, post something online. You can tell your board what to do by sending a set of instructions to the board's microcontroller. To do this, it uses the Arduino programming language (based on wiring) and the Arduino software (IDE), based on processing. Over the years, Arduino has been the brain of thousands of projects, from everyday objects to complex scientific instruments. A global community of creators (students, hobbyists, artists, programmers and professionals) has gathered around this open source platform, their contributions have been added to an incredible amount of accessible knowledge that can be of great help to beginners and experts. alike.

Arduino was born at the Ivrea Interaction Design Institute as an easy tool for rapid prototyping, aimed at students with no experience in electronics and programming. As soon as it reached a broader community, the Arduino board began to change to adapt to new needs and challenges, differentiating its offering from 8-bit single boards to products for IoT applications, portable devices, 3D printing and integrated environments. All Arduino boards are completely open source, allowing users to build them independently and eventually adapt them to their particular needs. The software is also open source and is growing through contributions from users around the world.

9.3.1 Important Features of Arduino Board

Arduino has been used in thousands of different projects and applications. The Arduino software is easy to use for beginners, but flexible enough for advanced users. It runs on Mac, Windows and Linux. Teachers and students use it to build inexpensive scientific instruments, to test the principles of chemistry and physics, or to get started with programming and robotics. Designers and architects build interactive prototypes, musicians and artists use it for installations and to experiment with new musical instruments. Creators, of course, use it to build many of the projects showcased at the Maker Fair, for example. Arduino is a key tool for learning new things. Anyone - kids, fans, artists, programmers - can start playing by following the step-by-step instructions in a kit or sharing ideas online with other members of the Arduino community. There are many other microcontrollers and microcontroller platforms available for physical computing. Parallax Basic Stamp, Netmedia's BX-24, Phidgets, MIT's Handyboard and many others offer similar functionality. All of these tools take the messy details of microcontroller programming and wrap it up in an easy-to-use package. Arduino also simplifies the process of working with microcontrollers, but offers some advantage for interested teachers, students, and hobbyists over other systems:

- Inexpensive: Arduino boards are relatively cheap compared to other microcontroller platforms. The least expensive version of the Arduino module can be assembled by hand, and even pre-assembled Arduino modules cost less than $ 50
- Cross-platform: Arduino software (IDE) runs on Windows, Macintosh OSX and Linux operating systems. Most microcontroller systems are limited to Windows.
- Simple and clear programming environment: Arduino software (IDE) is easy for beginners to use, but flexible enough for advanced users to take advantage of as well. For teachers, it is conveniently based on the processing programming environment, so students learning to program in that environment will be familiar with how the Arduino IDE works.
- Open source and extensible software: Arduino software is released as open source tools, available for extension by experienced programmers. The language can be extended through the C ++ libraries, and people who want to understand the technical details can take the

Arduino leap to the AVR C programming language on which it is based. Similarly, you can add the AVR-C code directly to your Arduino programs if you like.
- Open source and extensible hardware: Arduino board plans are released under a Creative Commons license, so experienced circuit designers can make their own version of the module, extend and upgrade it. Even relatively inexp users

Make connected devices easily with ARDUINO NANO 33 IoT products and open your creativity with the opportunities of the world wide web.

Figure 9.2: ARDUINO NANO 33 IoT

9.4 Raspberry Pi based IoT Projects

Raspberry Pi is a miniature card-sized computer that can perform all the functions we do with our desktop. Raspberry Pi was originally designed for educational purposes, but is now even being used for commercial activities. The board was inspired and built on BBC Micro technology from 1981.

Figure 9.3: Raspberry Pi Board

Eben Upton is the man behind this innovative hardware technology. He designed this platform to create a low-cost electronic device that would help enthusiasts and amateurs improve their programming skills and hardware understanding with ease.

Raspberry Pi serves as a pocket desktop computer that can be used as a media center, game consoles, or even as a router. In addition to these applications, Raspberry Pi is also used to develop many interesting projects. From mobile phones, tablets, robots, smart devices to a smart camera, everything can be developed on this platform. Now, this powerful Linux computer is being used to build many IoT projects that can transform any device more intelligently. Since IoT technology requires a microcontroller to process data, WiFi integration to send data to the cloud, and actuators to control operations, innovative minds around the world prefer to use Raspberry Pi to develop IoT projects. Since Raspberry Pi can perform all of the above mentioned operations smoothly without any external integration, it serves as a compression platform for building IoT projects. Some of the features of Raspberry Pi that make it an effective platform to build IoT projects are:
- A 1.4GHz 64-core 64-bit ARM Cortex-A53 CPU
- Dual Band 802.11ac Wireless LAN and Bluetooth 4.2
- Faster Ethernet (Gigabit Ethernet over USB 2.0)
- Power over Ethernet support
- Low energy consumption
- Enhanced PXE network and USB storage mass boot

IOT is changing the way we live. With more and more devices connected to the Internet, the demand for skills in this area is very high. You can learn IOT only by creating projects: get your hands on the sensors and actuators, configure the network, and collect and analyze the data sent by the sensors. In this project, you will learn how to build IOT projects using the most popular Raspberry Pi board. Using a Raspberry Pi computer and a DHT sensor, you will develop an electronic device that transmits temperature and humidity data over the Internet. By building this project, you will learn about:
- IoT - Concepts and applications
- Raspberry Pi development platform
- Data analysis with cloud platform

9.4.1 Raspberry Pi IoT based Smart Energy Monitor

Energy monitors, whether they cover the entire department or are deployed to monitor just one appliance, give you a way to track your usage and make necessary adjustments. While they are increasingly available on the market, the manufacturer in me still feels that it will be a great idea to build a DIY version that can be tailored to meet specific personal requirements. As such, for today's tutorial, we will create a Raspberry Pi power consumption monitor capable of obtaining power consumption and uploading it to Adafruit.io.

Raspberry Pi Smart Energy Meter Block Diagram

A block diagram showing how the system works is shown below.

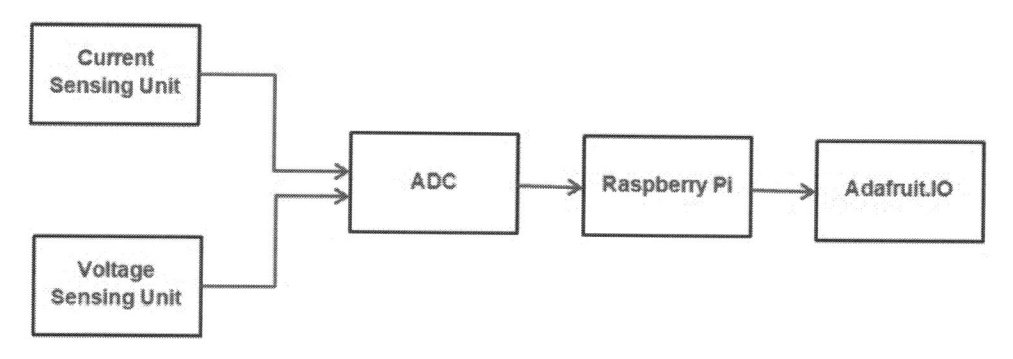

Figure 9.4: Raspberry Pi Smart Energy Meter Block Diagram

To choose the units one after another;

Current Detection Unit: The current detection unit is made up of the SCT -013 current sensor that can measure up to 100 A, depending on the version you buy. The sensor transforms the current that passes through the cable in which it is held into a small current that is then fed to the ADC through a network of voltage dividers.

Voltage Detection Unit: Although I couldn't get my hands on a voltage sensor module, we will build a DIY, transformerless voltage sensor that measures voltage using the voltage dividers principle. The DIY voltage sensor involves the stage of the voltage divider where the high voltage is transformed to a value suitable for input to the ADC.

Processing Unit: The processing unit comprises the ADC and the Raspberry pi. The ADC takes the analog signal and sends it to the raspberry pi, which then calculates the exact amount of energy consumed and sends it to a designated device cloud. For the purposes of this tutorial, we will use Adafruit.io as our Device Cloud.

Required Components
The following components are required to build this project;

- Raspberry Pi 3 or 4 (the process should be the same for the RPI2 with a WiFi Dongle)
- ADS1115 16bit I2C ADC
- YHDC SCT-013-000
- 2.5A 5V MicroUSB Power Adapter
- 2W 10K Resistor (1)
- 1/2W 10K Resistor (2)
- 33ohms Resistor (1)
- 2W 3.3k Resistor (1)
- IN4007 Diode (4)
- 3.6v Zener Diode (1)
- 10k Potentiometer(or Preset) (1)
- 50v 1uf Capacitor
- 50v 10uf Capacitor (2)
- BreadBoard
- Jumper Wire
- Other Accessories for Raspberry Pi's Use.

Preparing the Pi

Before we start wiring the components and coding, there are a few simple tasks that we need to do on the raspberry pi to make sure we are ready to go.

Step 1: Enable Pi I2C

The core of the current project is not just the raspberry pi but the I2C-based ADS1115 16-bit ADC. The ADC allows us to connect analog sensors to the Raspberry Pi, since the Pi does not have a built-in ADC. It takes the data through its own ADC and forwards it to the raspberry pi via I2C. As such, we must enable I2C communication on the Pi so that you can communicate with it. Pi's I2C bus can be enabled or disabled through the raspberry pi configuration page. To start it, click on the Pi icon on the desktop and select preferences, followed by Raspberry pi settings.

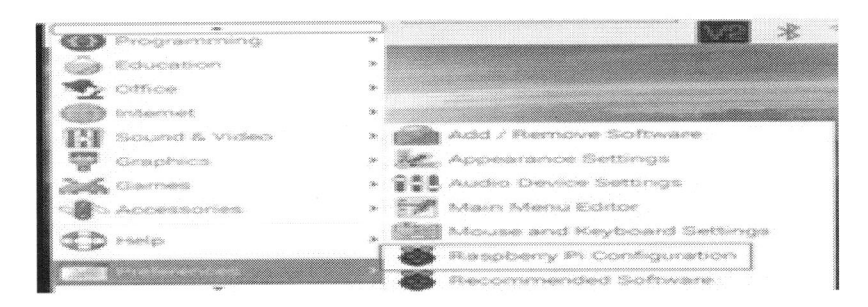

Figure 9.5: Raspberry pi configuration

This should open the configuration page. Check the I2C-enabled radio button and click OK to save it and restart the Pi to make the changes.

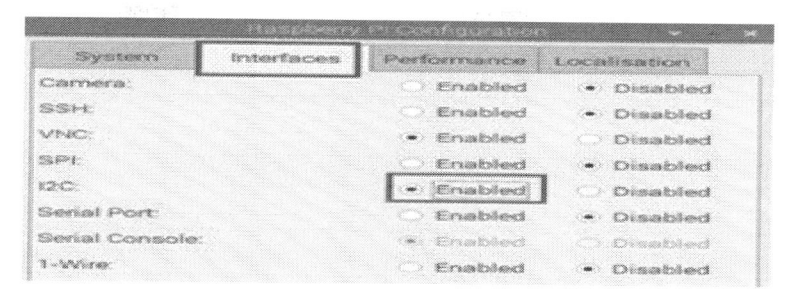

Figure 9.6: Enabled Radio Button for the I2C

If you are running the Pi in headless mode, the Raspbian configuration page can be accessed by running sudo raspi-config.

Step 2: Install Adafruit ADS11xx library
The second thing we should do is install the ADS11xx Python library that contains functions and routines that make it easy for us to write a Python script to get ADC values. Follow the steps below to do this.
1. Update your pi by running; sudo apt-get update followed by sudo apt-get upgrade this will update the pi ensuring there are no compatibility issues for any new software you choose to install.
2. Then run the command cd ~ to make sure you are in the home directory.
3. Next, install the essential build elements by running; sudo apt-get install build-essential python-dev python-smbus git
4. Next, clone the Adafruit git folder containing the ADS library by running; git clone https://github.com/adafruit/Adafruit_Python_ADS1x15.git
5. Change to the directory of the cloned file and run the configuration file using; cd Adafruit_Python_ADS1x1z followed by sudo python setup.py install
Once this is done, the installation should be complete.

You can test the library installation by connecting the ADS1115 as shown in the schematics section below and first run the sample code that came with the library, changing to your folder using; cd examples and run the example using; python simpletest.py

Step 3: Install Adafruit.IO Python Module

As mentioned during the presentations, we will post readings from the voltage and current sensors in Adafruit IO Cloud from where you can see it from all over the world or connect with IFTTT to perform the actions you want.

The Python Adafruit.IO module contains subroutines and functions that we will take advantage of to easily transmit data to the cloud. Follow the steps below to install the module.

1. Run cd ~ to return to the home directory.
2. Then run the command; sudo pip3 installs adafruit-io. You must install the Adafruit IO python module.

Step 4: Set up your Adafruit.io account

To use Adafruit IO you will definitely need to create an account first and get an AIO key. This AIO key along with your username will be used by your Python script to access the Adafruit IO cloud service. To create an account, visit; https://io.adafruit.com/, click on the Start for free button and complete all the required parameters. With the registration complete, you should see the View AIO key button on the right of your home page.

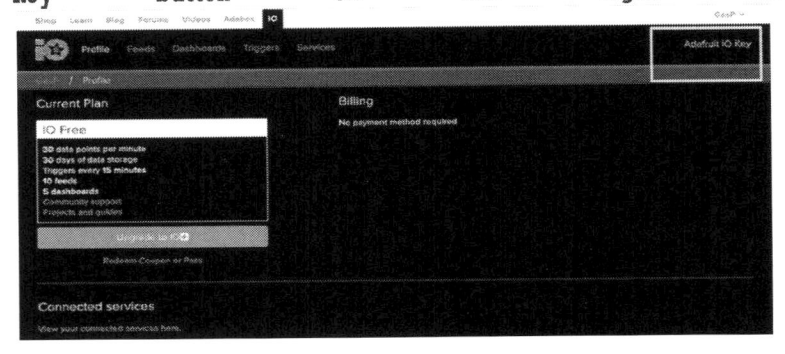

Figure 9.7: Setup Your Adafruit.io Account

With the key copied, we are ready to go. However, to facilitate the process of sending data to the cloud service, you can also create the sources to which the data will be sent. (You can find more information about what AIO fonts are here.) Since we will basically send the power consumption, we will create a power supply. To create a feed, click "feeds" at the top of the AIO page and click add new feed. Give it the name you want, but to keep things simple, I'll call it power consumption. You can also decide to create voltage and current sources and adapt the code to publish data. With all of this in place, we are now ready to start building the project.

Pi Energy Meter Circuit Diagram

The schemes for the Raspberry Pi energy monitor project are relatively complex and involves connecting to an AC voltage as mentioned above, be sure to take all necessary precautions to avoid electric shock. If you are not familiar with the safe handling of AC voltages, let the joy of implementing this on a breadboard without powering it be satisfactory. The schemes involve connecting the voltage and current sensor unit to the ADC which then sends the data from the sensors to the Raspberry Pi. To make the connections easier to follow, the diagrams of each unit present themselves.

Current Sensor Schematic

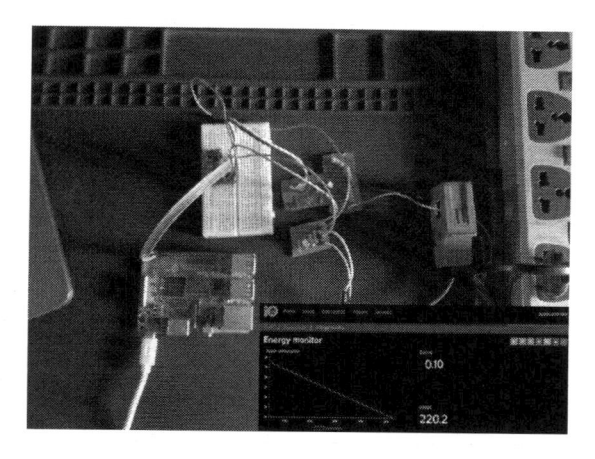

Figure 9.8: Current Sensor Schematic

Processing Unit Schematics

Connect everything together with the ADC (ADS1115) connected to the raspberry pi and the output of the current and voltage sensors connected to pins A0 and A1 of the ADS1115 respectively.

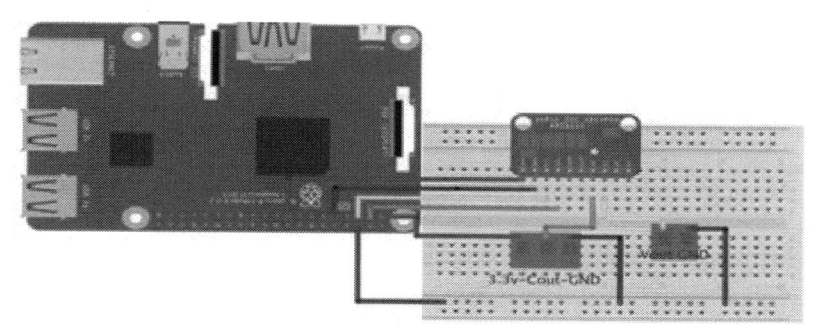

Figure 9.9: Processing Unit Schematics

Python Code for Pi Energy Meter

As usual with our raspberry pi projects, we will develop the code for the project using python. Click on the raspberry pi icon on the desktop, select the schedule and run the version of python you want to use. I will use Python 3 and some of the functions in python 3 may not work for python 2.7. Therefore, it may be necessary to make some significant code change if you want to use Python 2.7. I'll break the code down into little snippets and share the entire code with you at the end. The algorithm behind the code is simple. Our Python script queries ADS1115 (over I2C) for voltage and current readings. The received analog value is received, sampled, and the root mean square value of the voltage and current is obtained. The power in kilowatts is calculated and sent to the Adafruit IO power after specific intervals. We start the script including all the libraries that we will be using. This includes built-in libraries like the Time and Math Library and the other libraries we installed earlier.

```
import time
import Adafruit_ADS1x15
from Adafruit_IO import *
import math
```

Next, we create an instance of the ADS1115 library that will be used to address the physical ADC in the future.

```
# Create an ADS1115 ADC instance (16 bit).
adc1 = Adafruit_ADS1x15.ADS1115 ()
```

Then provide your adafruit IO username and "AIO" key.

```
username = 'enter your username in these quotes'
AIO_KEY = 'your aio key'
aio = Client (username, AIO_KEY)
```

Please keep the key safe. It can be used to access your adafruit io account without your permission. Next, we create some variables like the gain for the ADC, the number of samples we want and set the rounding which is definitely not critical.

```
GAIN = 1 # consult the documentation of ads1015 / 1115 for the potential values.
samples = 200 # number of samples taken from ads1115
places = int (2) # rounding set
```

Next, we create a while loop to monitor current and voltage and send the data to Adafruit io at intervals. The while loop begins by setting all variables to zero.

```
while true:
    # reset variables
    account = int (0)
    datai = []
    datav = []
    maxIValue = 0 #max current value within the sample
    maxVValue = 0 # maximum voltage value within the sample
    IrmsA0 = 0 #root mean square current
    VrmsA1 = 0 # root mean square stress
    ampsA0 = 0 # peak current
    voltsA1 = 0 #voltage
```

```
kilowatts = float (0)
```

Since we are working with AC circuits, SCT-013's output and voltage sensor will be a sine wave, therefore, to calculate the current and voltage of the sine wave, we will need to get the maximum values. To obtain the maximum values, we will take voltage and current samples (200 samples), and we will find the highest values (maximum values).

```
for range count (samples):
    datai.insert (count, (abs (adc1.read_adc (0, gain = GAIN))))
    datav.insert (count, (abs (adc1.read_adc (1, gain = GAIN))))
    # see if you have a new maxValue
    print (data [count])
    if datai [account]> maxIValue:
        maxIValue = datai [account]
    if datav [count]> maxVValue:
        maxVValue = datav [account]
```

Next, we standardize the values by converting the ADC values to the actual value, after which we use the root-mean-square quadratic equation to find the RMS voltage and current.

```
# calculate current using sampled data
    # the sct-013 that is used is calibrated for an output of 1000 mV at 30 A
    IrmsA0 = float (maxIValue / float (2047) * 30)
    IrmsA0 = round (IrmsA0, places)
    ampsA0 = IrmsA0 / math.sqrt (2)
    ampsA0 = round (ampsA0, places)
    # Calculate voltage
    VrmsA1 = float (maxVValue * 1100 / float (2047))
    VrmsA1 = round (VrmsA1, places)
    voltsA1 = VrmsA1 / math.sqrt (2)
    voltsA1 = round (voltsA1, places)
    print ('Voltage: {0}'. format (volts A1))
    print ('Current: {0}'. format (ampsA0))
```

Once this is done, the power is calculated and the data is published on adafruit.io

```
# computing power
power = round (amps A0 * volts A1, places)
print ('Power: {0}'. format (power))
```

```
#post data to adafruit.io
EnergyUsage = aio.feeds ('EnergyUsage')
aio.send_data ('EnergyUsage ', power)
```

For free accounts, adafruit requires that there be a time delay between requests or data upload.

```
# Wait before repeating the cycle
time.sleep (0)
```

To go further, you can create a dashboard in adafruit.io and add a graphical component so you can get a graphical view of the data as shown in the image below.

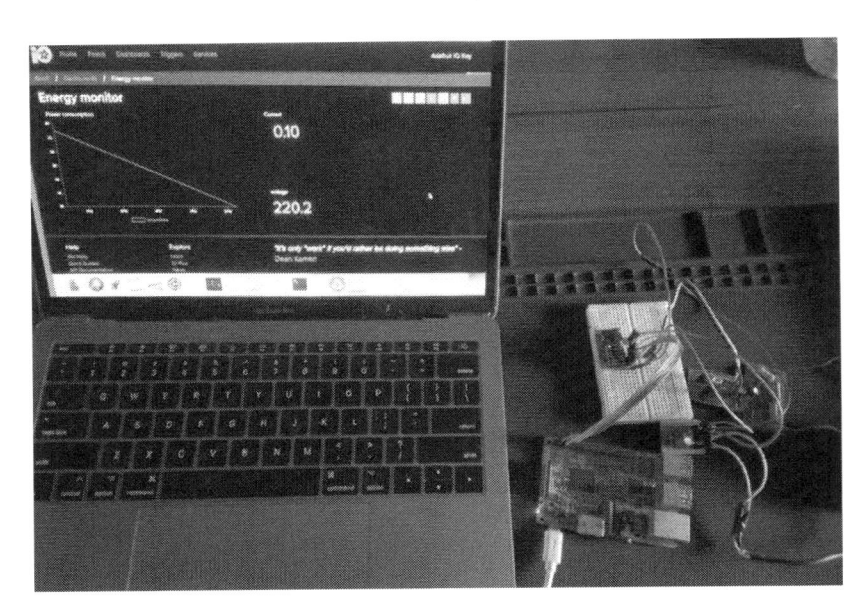

Figure 9.10: Graphical view of the Energy Meter Data

9.4.2 IoT based Smart Parking System

Components Required
- ESP8266 NodeMCU
- Ultrasonic Sensor
- DC Servo Motor
- IR Sensors
- 16x2 i2c LCD Display
- Jumpers

With the growing popularity of Smart Cities, there is always a demand for smart solutions for every domain. The IoT has enabled the possibility of Smart Cities with its Internet control function. A person can control the devices installed in their home or office from anywhere in the world simply using a smartphone or any device connected to the Internet. There are multiple domains in a smart city, and Smart Parking is one of the popular domains in the smart city. The Smart Parking industry has seen a number of innovations, such as the Smart Parking Management System, Smart Gate Control, Smart Cameras that can detect vehicle types, ANPR (Automatic License Plate Recognition), Smart Payment System, Entry System smart and many more. Today a similar approach will be followed and a smart parking solution will be built that will use an ultrasonic sensor to detect the presence of the vehicle and activate the door to automatically open or close. The ESP8266 NodeMCU will be used here as the primary controller to control all peripherals connected to it. In this intelligent IoT parking system, we will send data to the web server to search for the availability of space for vehicle parking. Here we are using firebase as Iot database to get the parking availability data. For this, we need to find the Firebase host address and secret key for authorization. If you already know how to use firebase with NodeMCU, then you can go ahead; otherwise, you must first learn how to use Google Firebase Console with ESP8266 NodeMCU to get the host address and secret key.

Circuit Diagram

The circuit diagram for this IoT based vehicle parking system is given below. They are two IR sensors, two servo motors, an ultrasonic sensor and a 16x2 LCD. Here the ESP8266 will control the entire process and also send the parking availability information to Google Firebase so that it can be monitored from anywhere in the world via the Internet. Two IR sensors are used at the entrance and exit doors to detect the presence of the car and automatically open or close the door. The IR sensor is used to detect any object by sending and receiving IR rays. Find out more about the IR sensor here. Two servos will act as the entry and exit door and will rotate to open or close the door. Finally, an ultrasonic sensor is used to detect if the parking space is available or occupied and send the data to ESP8266 accordingly.

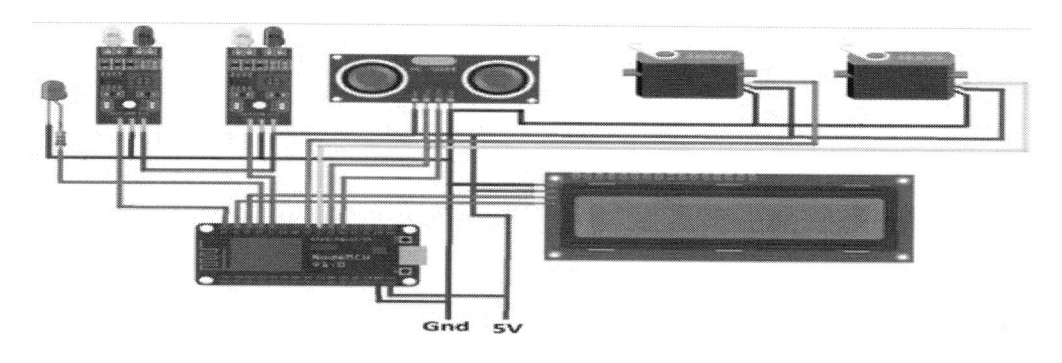

Figure 9.11: Circuit Diagram

Programming ESP8266 NodeMCU for Smart Parking Solution

To program NodeMCU, simply connect the NodeMCU to the computer with a micro USB cable and open Arduino IDE. Libraries are necessary for I2C Display and Servo Motor. The LCD screen will show the availability of parking spaces and the servo motors will be used to open and close the entrance and exit doors. The Wire.h library will be used to connect the LCD screen in the i2c protocol. The pins for I2C on the ESP8266 NodeMCU are D1 (SCL) and D2 (SDA). The database used here will be Firebase, so here we also include the library (FirebaseArduino.h) for it.

#include <ESP8266WiFi.h>
#include <Servo.h>
#include <LiquidCrystal_I2C.h>
#include <Wire.h>
#include <FirebaseArduino.h>

Then include the Firebase credentials obtained from Google Firebase. These will include the hostname that contains the name of your project and a secret key. To find these values, follow the previous tutorial on Firebase.

#define FIREBASE_HOST "smart-parking-7f5b6.firebaseio.com"
#define FIREBASE_AUTH "suAkUQ4wXRPW7nA0zJQVsx3H2LmeBDPGmfTMBHCT"
Include Wi-Fi credentials like WiFi SSID and password.
#define WIFI_SSID "CircuitDigest"
#define WIFI_PASSWORD "circuitdigest101"

Initialize the I2C LCD with the device address (here is 0x27) and the LCD type. Also include servo motors for the entrance and exit doors.

LiquidCrystal_I2C lcd (0x27, 16, 2);
Servo myservo;
Servo myservo1;
Initiate I2C communication for I2C LCD.
Wire.begin (D2, D1);

Connect the input and output servo motor to pins D5, D6 of the NodeMCU.

myservo.attach (D6);
myservos.attach (D5);

Select the trigger pin of the ultrasonic sensor as the output and the echo pin as the input. The ultrasonic sensor will be used to detect the availability of the parking place. If the car has taken up the space, then it will shine; otherwise it will not shine.

```
pinMode (TRIG, OUTPUT);
pinMode (ECHO, INPUT);
```

The two pins D0 and D4 of the NodeMCU are used to take the reading from the IR sensor. The IR sensor will act as an entry and exit door sensor. This will detect the presence of the car.

```
pinMode (carExited, INPUT);
pinMode (carEnter, INPUT);
```

Connect to WiFi and wait a bit until it connects.

```
WiFi.begin (WIFI_SSID, WIFI_PASSWORD);
Serial.print ("Connection to");
Serial.print (WIFI_SSID);
while (WiFi.status ()! = WL_CONNECTED) {
  Serial.print (".");
  delay (500);
}
```

Start the connection to Firebase with Host and secret key as credentials.

```
Firebase.begin (FIREBASE_HOST, FIREBASE_AUTH);
```

Start i2c 16x2 LCD and set the cursor position to the 0th row 0th column.
```
lcd.begin ();
lcd.setCursor (0, 0);
```

Take the distance from the ultrasonic sensor. This will be used to detect the presence of the vehicle at the particular location. Send the 2 microsecond pulse first, and then read the received pulse. Then convert it to 'cm'. Find out more about distance measurement with an ultrasonic sensor here

```
digitalWrite (TRIG, LOW);
delayMicroseconds (2);
digitalWrite (TRIG, HIGH);
delayMicroseconds (10);
digitalWrite (TRIG, LOW);
```

```
duration = pulseIn (ECHO, HIGH);
distance = (duration / 2) / 29.1;
```

Digitally read the IR sensor pin as input sensor and check if it is high. If it is high, increase the input count and print it on a 16x2 LCD screen and also on a serial monitor.

```
int carEntry = digitalRead (carEnter);
if (carEntry == HIGH) {
  countYes ++;
  Serial.print ("Auto entered ="); Serial.println (countSes);
  lcd.setCursor (0, 1);
  lcd.print ("Auto entered");
```

Also move the angle of the servomotor to open the entrance door. Change the angle according to your use case.

```
  for (pos = 140; pos> = 45; pos - = 1) {
    myservos.write (pos);
    delay (5);
  }
  delay (2000);
  for (pos = 45; pos <= 140; pos + = 1) {
    // in steps of 1 degree
    myservos.write (pos);
    delay (5);
  }
```

And send the read to firebase using the pushString function of the Firebase library.

```
Firebase.pushString ("/ Parking Status /", fireAvailable);
```

Perform the steps similar to the above for the output IR sensor and the output servo motor.

```
int carExit = digitalRead (carExited);
if (carExit == HIGH) {
  account Yes;
  Serial.print ("Car Exited ="); Serial.println (countSes);
  lcd.setCursor (0, 1);
  lcd.print ("Car exited");
  for (pos1 = 140; pos1> = 45; pos1 - = 1) {
    myservo.write (pos1);
```

```
    delay (5);
  }
  delay (2000);

  for (pos1 = 45; pos1 <= 140; pos1 + = 1) {
    // in steps of 1 degree
    myservo.write (pos1);
    delay (5);
  }
  Firebase.pushString ("/ Parking Status /", fireAvailable);
  lcd.clear ();
}
```

Check if the car has reached the parking place and if it has arrived, then the pilot light will give you the signal that the place is full.

```
  yes (distance <6) {
       Serial.println ("Busy");
     digitalWrite (led, HIGH);
  }
```

Otherwise, show that the place is available.

```
  if (distance> 6) {
       Serial.println ("Available");
     digitalWrite (led, LOW);
  }
```

Calculate the total empty space inside the parking lot and save it in the chain to send the data to firebase.

```
Empty = allSpace - countYes;
  Available = String ("Available =") + String (empty) + String ("/") + String (allSpace);
  fireAvailable = String ("Available =") + String (Empty) + String ("/") + String (allSpace);
```

Also print the data on the i2C LCD screen.
```
  lcd.setCursor (0, 0);
  lcd.print (available);
```

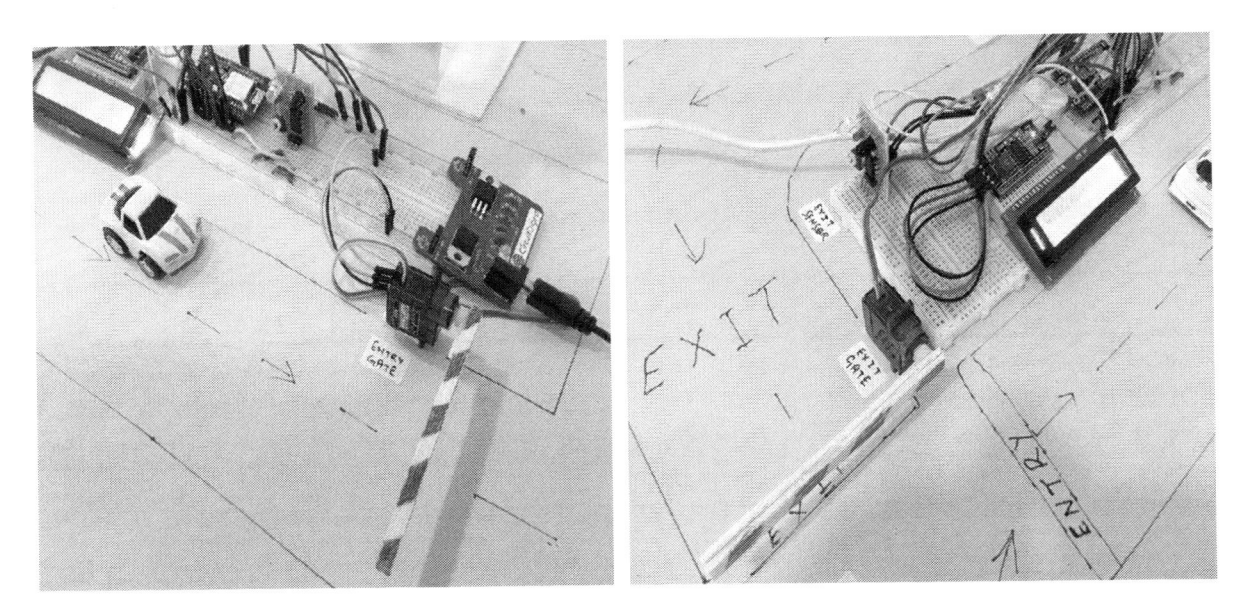

Figure 9.12: Smart Parking Solution

Here's how you can track availability of online parking in Firebase as shown in the following snapshot:

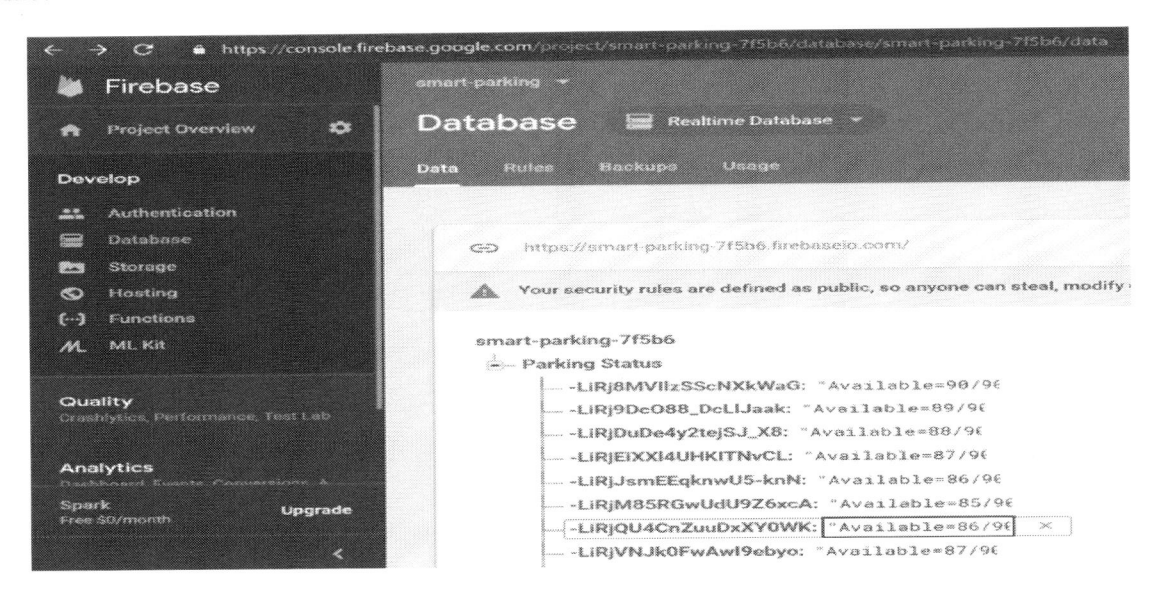

Figure 9.13: Availability of parking can be tracked online on Firebase

This ends the entire smart parking system using the ESP8266 NodeMCU module and different peripherals. You can use other sensors also as a replacement for the ultrasonic and IR sensor. There is a wide application of Smart Parking System and different products can be added to make it smarter.

9.4.3 IoT based Smart Irrigation System

Most farmers use large portions of agricultural land and it becomes very difficult to reach and track every corner of the large land. At some point there is the possibility of uneven water splashes. This results in poor quality crops that lead to financial losses. In this scenario, the Smart Irrigation System using the latest IoT technology is useful and makes farming easier. The smart irrigation system has a wide scope to automate the entire irrigation system. Here we are building an IoT based irrigation system using the ESP8266 Node MCU module and the DHT11 sensor. Not only will it automatically irrigate the water based on the level of moisture in the soil, but it will also send the Data to the ThingSpeak Server to track terrain conditions. The system will consist of a water pump that will be used to spray water on the ground, depending on the environmental conditions of the land, such as humidity, temperature, and humidity. Before starting, it is important to note that different crops require different soil moisture, temperature, and humidity conditions. So in this tutorial we are using a crop that will require a soil moisture of about 50-55%. So when the soil loses its humidity to less than 50%, the motor pump will automatically turn on to spray the water and continue to spray the water until the humidity reaches 55% and then the pump will shut off. The sensor data will be sent to the ThingSpeak server at a defined time interval so that it can be monitored from anywhere in the world.

Components Required
- NodeMCU ESP8266
- Soil Moisture Sensor Module
- Water Pump Module
- Relay Module
- DHT11
- Connecting Wires

Circuit Diagram

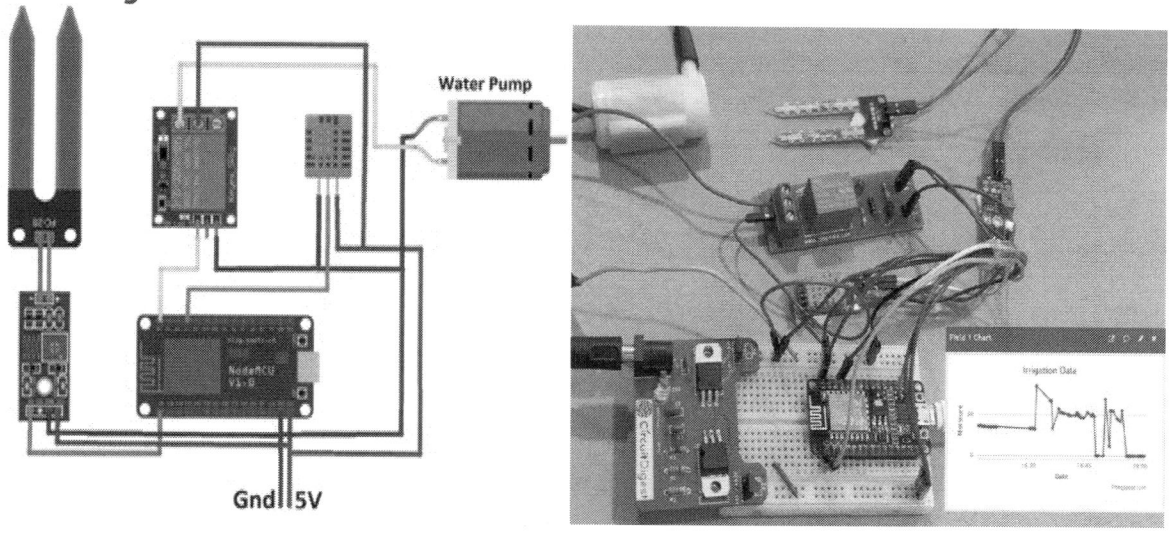

Figure 9.14: Circuit Diagram of IoT based Smart Irrigation System

Programming ESP8266 Node MCU for Automatic Irrigation System

To program the ESP8266 Node MCU module, only the DHT11 sensor library is used as the external library. The humidity sensor provides an analog output that can be read through the ESP8266 Node MCU A0 analog pin. Since the node MCU cannot provide an output voltage above 3.3V from its GPIO, we are using a relay module to drive the 5V motor pump. In addition, the humidity sensor and DHT11 sensor are powered from an external 5V power supply. The complete code with a working video is at the end of this tutorial, here we explain the program to understand the project workflow.

Start by including the required library

```
#include <DHT.h>
#include <ESP8266WiFi.h>
```

Since we are using the ThingSpeak server, the API key is necessary in order to communicate with the server. To find out how we can get the ThingSpeak API key, you can visit the previous article on Live Temperature and Humidity Monitoring on ThingSpeak.

```
String apiKey = "X5AQ445IKMBYW31H
const char * server = "api.thingspeak.com";
```

The next step is to type in the Wi-Fi credentials, such as SSID and password.

```
const char * ssid = "CircuitDigest";
const char * pass = "xxxxxxxxxxx";
```

Define the pin of the DHT sensor where the DHT is connected and choose the type of DHT.

```
#define DHTPIN D3
DHT dht (DHTPIN, DHT11);
```

The humidity sensor output is connected to Pin A0 of ESP8266 NodeMCU. And the motor pin is connected to D0 of NodeMCU.

```
constant int humidityPin = A0;
const int motorPin = D0;
```

We will use the millis () function to send the data after each time interval defined here, it is 10 seconds. Delay () is avoided since it stops the program during a defined delay in which the microcontroller cannot perform other tasks. Learn more about the difference between delay () and millis () here.

```
long unsigned interval = 10000;
unsigned long previousMillis = 0;
```

Set the motor pin as output and turn off the motor initially. Start reading the DHT11 sensor.

```
pinMode (motorPin, OUTPUT);
digitalWrite (motorPin, LOW); // keep the engine off initially
dht.begin ();
```

Try connecting the Wi-Fi with the given SSID and Password and wait for the Wi-Fi to connect and if connected go to the next steps.

```
WiFi.begin (ssid, pass);
  while (WiFi.status ()! = WL_CONNECTED)
  {
    delay (500);
    Serial.print (".");
  }
  Serial.println ("");
  Serial.println ("WiFi connected");
}
```

Define the current program start time and save it in a variable to compare with the elapsed time.

```
unsigned long currentMillis = millis ();
```

Read the temperature and humidity data and save it in variables.

```
float h = dht.readHumidity ();
float t = dht.readTemperature ();
```

If the DHT is connected and the ESP8266 NodeMCU can read the readings, continue with the next step or return from here to check again.

```
if (isnan (h) || isnan (t))
  {
    Serial.println ("Error reading from DHT sensor!");
```

```
    Return;
}
```
Read the humidity reading from the sensor and print the reading.

```
Moisture percentage = (100.00 - ((analogRead (idityPin) / 1023.00) * 100.00));
    Serial.print ("The soil moisture is =");
    Serial.print (percentage of humidity);
    Serial.println ("%");
```

If the humidity reading is within the required soil moisture range, keep the pump turned off or if it goes beyond the required humidity, turn the pump on.

```
if (humidity percentage <50) {
    digitalWrite (motorPin, HIGH);
}
  if (humidity percentage> 50 && humidity percentage <55) {
    digitalWrite (motorPin, HIGH);
}
 if (humidity percentage> 56) {
    digitalWrite (motorPin, LOW);
}
```

Now every 10 seconds call the sendThingspeak () function to send the humidity, temperature and humidity data to the ThingSpeak server.

```
    if ((unsigned long) (currentMillis - previousMillis)> = interval) {
    sendThingspeak ();
    previousMillis = millis ();
    client.stop ();
}
```

In the sendThingspeak () function we check if the system is connected to the server and, if so, we prepare a string where humidity, temperature, humidity reading is written and this string will be sent to the ThingSpeak server along with the API key and the address of the server.

```
if (client.connect (server, 80))
    {
        String postStr = apiKey;
        postStr + = "& field1 =";
```

```
postStr + = String (moisture percentage);
postStr + = "& field2 =";
postStr + = String (t);
postStr + = "& field3 =";
postStr + = String (h);
postStr + = "\ r \ n \ r \ n";
```

Finally, the data is sent to the ThingSpeak server using the client.print () function that contains the API key, the server address, and the string that was prepared in the previous step.

```
client.print ("POST / HTTP update / 1.1 \ n");
client.print ("Host: api.thingspeak.com \ n");
client.print ("Connection: close \ n");
client.print ("X-THINGSPEAKAPIKEY:" + apiKey + "\ n");
client.print ("Content type: application / x-www-form-urlencoded \ n");
client.print ("Content-Length:");
client.print (postStr.length ());
client.print ("\ n \ n");
client.print (postStr);
```

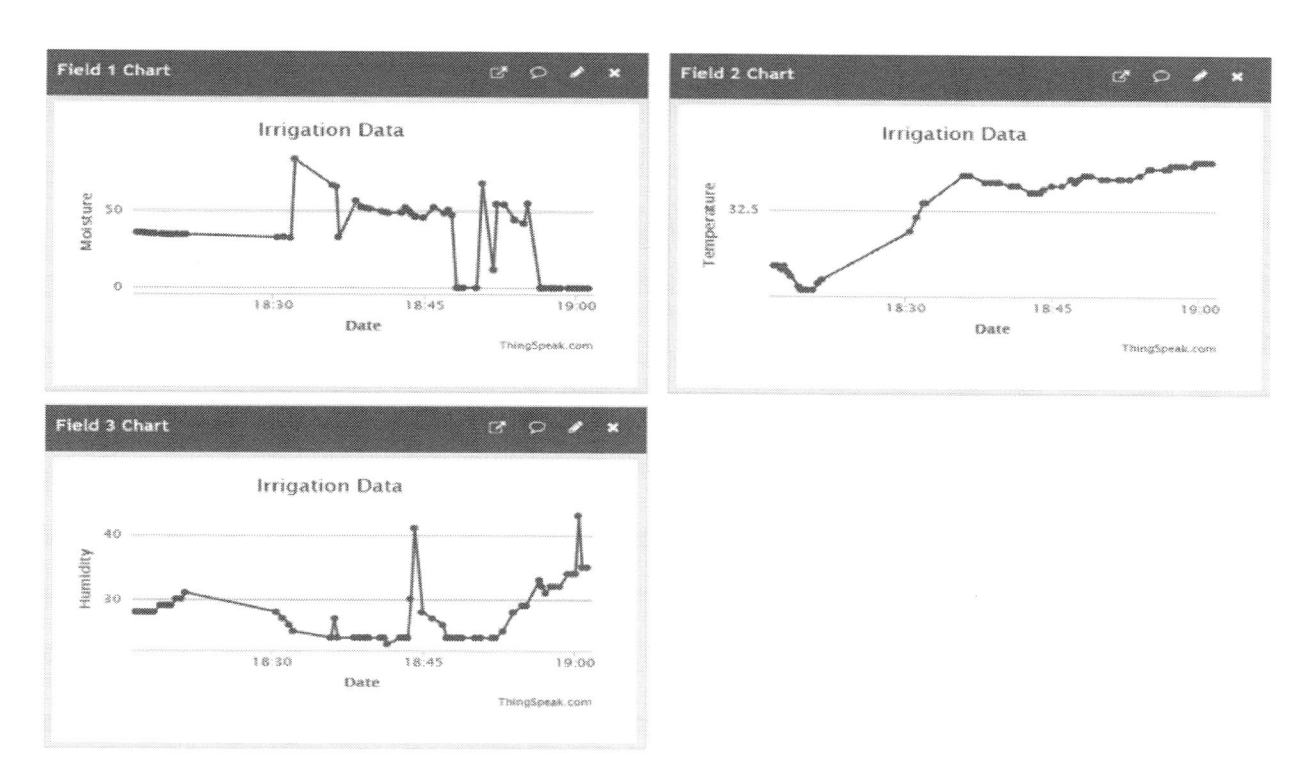

Figure 9.15: Plot of Irrigation Data

Chapter-10: Challenges and Future Scope of IoT

10.1 Challenges in IoT

The Internet of Things refers to interrelated devices that can transfer data over a network without the need for computing devices and human interaction. The Internet of Things is directly and indirectly associated with everyday lifestyle products around the world. The Internet of Things (IoT) is an advanced automation and analytics system that uses an ecosystem of networks, sensors, and actuators, big data analytics engine, cloud computing, and smart technologies to deliver services. IoT will interconnect all objects, living or non-living. It is difficult to differentiate and know the manifestation in IoT. First of all, what comes up in everyone's day to day are mobile phones, and the critical application that everyone uses is the video call or the video recorder. Here comes the but do you have a firewall in your home to protect your device from being attacked or hacked? Because there is a good chance that your personal information and data could be exposed to hackers or malicious users. Even if we have this threat, we can see that there are more IoT devices. People are still searching the online platform and offline platform for more advanced and better technology devices. In the end, we can say that we are experiencing an overgrowth of IoT devices.

Figure 10.1 : Challenges in IoT

In 2022 we can say that tons and tons more of devices connected to the Internet of Things are on the way. It is sometimes believed that major internet providers like Verizon, The AT & Ts are not yet ready to provide fiber connectivity. All this is because the infrastructure is sufficient to support it.

10.1.1 Major Challenges in IoT

If we talk about the driverless car, this device collects a certain amount of data that is huge and incredible. The first one here is the Petabytes information. This requires regular data processing and transfer over the network to the most central platform that will give the right decision. This directly indicates that this technology is not yet ready and safe. In 2020, there were a large number of data breaches by major companies, such as Facebook data breach, Google data breach, some airways and hotels have seen a major data breach . Therefore, companies became more alert about all of this, and we can expect security to be the main thing that companies will deal with.

10.2 Future Scope of IoT

Whether people are connected or not, but the opportunity will remain, with the massive increase in the use of IoT devices. This can lead to security and freedom issues that IoT devices add to.

10.2.1 Addiction to Tech Connections

Surveys suggest that the use of IoT-based devices will increase enormously. It will influence people during this and the next decade. There will be some of the magic things that will make people addictive to devices and people will not be able to resist leaving it. Convenience and benefits will keep people drawn to the devices. In the future, we can even expect people to choose connectivity over security. By connecting with society, friends and new technologies in comfort will become more critical. People will even begin to store all their information data about themselves and their families on these devices, and will begin to market security for their convenience. Users will begin to make all rational decisions based on their safety and protection. Children, adults, everyone will become addicted, and the house will become a great IoT device. While human beings will not be able to cope without all this.

Companies will begin to penalize company employees who are connected to the Internet or those who disconnect. Staying active in digital media and social media will earn rewards. However, full withdrawal will be very difficult and may be impossible. We can also hear stories where you could see things like people who tried to disconnect but failed. Although you will not be able to take off with all these devices. As you will be surrounded by all these things and you will not be able to avoid it. People will have the illusion of being separated, and it will not be true.

10.2.2 Increase in Internet Participants

There will be people who will try to disconnect from digital media and social media. Finally, as soon as they try to split twice as much, people will come and join the internet platform. It just doubles but can never go down. Smart TVs, smart phones, voice searches will become an essential requirement. Even disconnecting here means that those users are looking for a better option to connect again. Ultimately, it means that users always double rather than decrease.

10.2.3 Risk Mitigation and IoT Safer

The ingenuity of human beings will make them genuine users, in the same way, risk mitigation will decrease. The Internet of Things will be more reliable for users. However, the race for authenticity and security will be high and needed, but the level of security is guaranteed to increase dramatically and deliver the best results. All networked devices will be reliable like the power grid and will make the platform safe from wrong actions. However, at certain points it will occasionally go out, and that will be a disaster because people, of course, will survive. You will gain many technology-based and regulatory breakthroughs that will act as remedies and lessen all threats. As citizens of countries and people around the world become much more aware of security and safety on the Internet platform. This will increase pressure on the government if something goes wrong. It will make companies automatically push themselves and everything is safe for users.

10.2.4 Increase in Risk

Whether people are connected or not, but the opportunity will remain, with the massive increase in the use of IoT devices. This can lead to security and freedom issues that IoT devices add to.

Figure 10.2 :Threats in IoT

Threats can turn into horrifying attacks and all other acts that can be very violent. Physical attacks are public and people can see them. However in cyber attacks it will be in private and you will not know who the attacker is but the results are terrible. With the rise of IoT and security, concern will increase as user freedoms. You can say this will help you know where you are walking and light your way, or you can get your confidential and personal information. It will become the biggest challenge for the police, the government and the entire world. As we come to an end, the IoT or Internet of Things has made human life simple and comfortable. It has made people's lives very convenient. While, on the other hand, with the increasing use of the Internet of Things, security and safety have also increased. Therefore, we must be careful when providing details on the Internet platform. However, we can see that many necessary steps are being taken, but keeping your data safe with you is essential.

10.3 Future of the Internet of Things (IoT)

Kevin Ashton is the father of IoT (Internet of Things) which represents a system where the Internet is connected to the real world through generalized sensors. IoT has unlimited potential, orchestration, and deployment capabilities. More than millions of organizations have adopted one of the leading providers of cloud platforms. Large companies like Amazon, Google and IBM offer a wide range of services to collect IoT information for data analysis.

10.3.1 Internet of Things Market Size

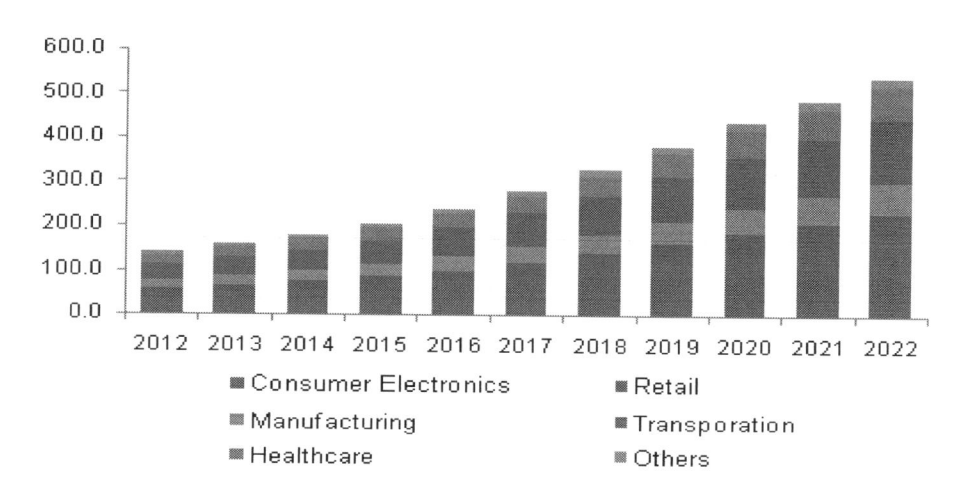

Figure 10.3: Market Size of IoT

10.3.2 The Future of IoT is AI

AI offers power to unlock the potential of IoT: Artificial Intelligence plays a vital role in IoT applications in startups that combine analytics capabilities based on machine learning. Machine learning is an AI technology that has the potential to detect anomalies and redundant data generated by smart sensors.

- **Operational efficiency and risk management:** AI enables better offerings to give a competitive advantage in business performance. There are numerous AI-connected applications that help companies predict the risk of a rapid response to manage cyber threats and financial losses.

- **Enabling new services and products:** To create products and implement new services, it is essential to improve artificial intelligence and machine learning.

- **The concept of IoT and machine learning is not new in this computing world:** Machine learning uses various learning techniques on historical data to make decisions. Decision making becomes easy if the amount of historical data is substantial. The future of the Internet of Things has the potential to be limitless. Approaches to the industrial

Internet will be stimulated through enhanced network activity, integrated artificial intelligence (AI), and the ability to deploy, automate, orchestrate, and secure diverse use cases on a large scale. The potential is not only to invest billions of devices simultaneously, but also to take advantage of the large volumes of actionable data that can automate different business processes. As IoT networks and platforms emerge to overcome these challenges, through increased capacity and artificial intelligence, service providers will move even closer to the IT and web-scale markets, opening up new sources of revenue.

- **Internet of Things by 2025:** Internet of Things devices such as machines and sensors are expected to generate 79.4 zettabytes of data in 2025, which IDC (International Data Corporation) predicts. Additionally, IoT will increase at a compound annual growth rate of 28.7% between 2020 and 2025.

- **The Internet of Things increases artificial intelligence:** We know that IoT refers to the device that transfers data over a network that produces incredible amounts of data, and many industries have no idea how to handle this amount of data. Furthermore, no company will ignore this data if it is linked to customers and their personal information.

- **5G - The Fuel IOT:** 5G Needs is Needed for the IoT or Individual Network for Billions of Applications. In a span of ten years, from 2020 to 2030, IoT devices will range from 75 billion to over 100 billion, and the move from 4G to 5G in terms of improving IoT is quite remarkable. The current 4G network can support up to 5,500 to 6,000 NB-IOT devices in a single cell. With a 5G network, a particular cell can control up to a million devices. Major Wireless rolled out 5G networks in 2019. 5G promises greater speed and the ability to connect more smart devices simultaneously. Faster networks mean that the data collected by your smart devices will be further analyzed and managed. It will also drive innovation in companies that make IoT devices and increase consumer demand for new products.

- **There are an estimated 21B more IoT devices by 2025** - let's take a quick look at the evolution of IoT devices. In 2016, there were more than 4.7B devices connected to the Internet, according to the IoT analysis. In 2021 and 2025, it is expected to increase to 11.6B and 21B devices, respectively.

- **More cities will get smart:** Plus, consumers, cities, and businesses will quickly adopt smart technologies to save time, money, and energy. It will help cities automate, manage, and collect data through visitor kiosks, video camera surveillance systems. Taxis, bicycle rental stations and many more.

- **Cars will get smarter:** the merger of 5G and IoT will change the auto industry at a higher speed. The arrival of driverless cars and connected vehicles will facilitate the transportation system.

10.4 IoT Applications

Smart Life: By providing Social Security with another patient-driven model, imaginative models for managing an account and a fund, an effective and advantageous transmission of open administrations.

Smart City: by providing smart frame management, smart metering and network management, brilliant observation, safer and more mechanized transport, more vitality management frameworks

Smart Manufacturing: Works by inserting sensors into the solid during the pour and cure process. The sensors provide key data on solid quality and quality directly to the Smart Structures workstation. A Google search for the term "Internet of Things" reveals more than 280,000,000 results, thanks to the means that make the connection between portable home devices and the connected car. IOT is projected to be in the top 10 trends for the next 5 years.

Figure 10.4: Applications of IoT

Healthcare- Health technologists have been experimenting with IoT for over a decade and have developed some truly amazing devices. In addition to connecting patient tracking devices (EKG machines, for example) to the Internet, IoT functionality has even been added to some devices that are inserted directly into people during surgery, such as pacemakers.

Figure 10.5: Applications of IoT in Health Care

Home Security- In the sphere of home automation, IoT security devices are widely used. These devices include security alarms, door locks, surveillance cameras, and more. Homeowners can control these IoT security devices remotely from their smartphones or computers.

Chemical Plants- For hazardous areas like nuclear sites, IoT sensors are installed around job sites. The results of regular environmental scans are automatically uploaded to the cloud and, when dangerous levels are detected, can alert site administrators and initiate emergency protocols.

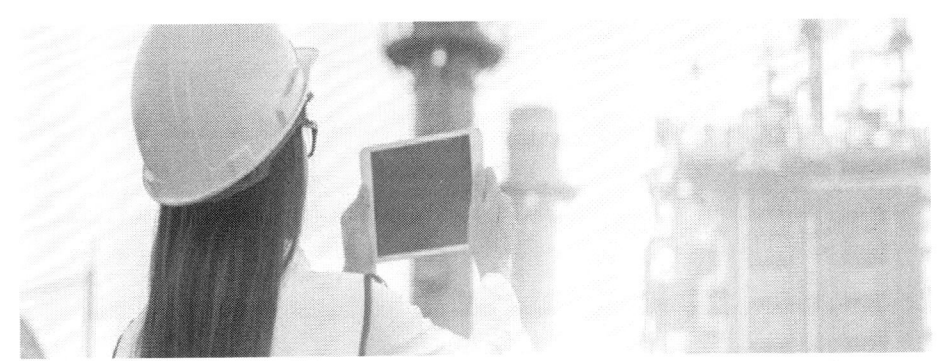

Figure 10.6: Applications of IoT in Chemical Plants

10.4.1 IoT Solutions over the Next Five Years

1. Businesses will be the main ones to adopt IoT solutions. IoT can improve your bottom line by 1) reducing operating costs; 2) increased productivity; and 3) expand into new markets or develop new product offerings.

2. Governments focused on increasing productivity, lowering costs and improving the quality of life of their citizens will be the second to adopt IoT ecosystems.

3. Therefore, the areas with IOT will be artificial intelligence, machine learning, and smart devices for: connected homes, smart cities, transportation, manufacturing, healthcare, utilities (smart grid), and agriculture applications.

4. Within IOT, machine learning and artificial intelligence will lead the trend.

AI / AML has allowed computers to compete or exceed human skills in many domains. Machines with the ability to perform complex jobs enable a new level of automation in all areas of business: auto answer, resume match, object detection, and pattern recognition.

Printed in Great Britain
by Amazon